PLANT PATHOLOGY

PLANT PATHOLOGY

STEPHEN BURCHETT
SARAH BURCHETT

Garland Science
Taylor & Francis Group
NEW YORK AND LONDON

Vice President: Denise Schanck
Assistant Editor: David Borrowdale
Production Editor: Daniela Amodeo
Illustrator: NovaTechset Ptv Ltd
Layout: NovaTechset Ptv Ltd
Cover Designer: Andrew Magee
Copyeditor: Josephine Hargreaves
Proofreader: Sally Huish
Indexer: Simon Yapp

Stephen Burchett, Lecturer in Ecology, School of Biological & Marine Sciences, Plymouth University. **Sarah Burchett**, Associate Lecturer, School of Geography, Earth and Environmental Sciences, Plymouth University.

Cover image: Powdery mildew on clematis bloom.

ISBN 978-0-8153-4483-4

Library of Congress Cataloging-in-Publication Data
Names: Burchett, Stephen, author. | Burchett, Sarah, author.
Title: Plant pathology / Stephen Burchett, Sarah Burchett.
Description: New York : Garland Science, Taylor & Francis Group, 2017. |
Includes bibliographical references and index.
Identifiers: LCCN 2017046379 | ISBN 9780815344834
Subjects: LCSH: Plant diseases.
Classification: LCC SB601 .B87 2017 | DDC 632--dc23
LC record available at https://lccn.loc.gov/2017046379

Published by Garland Science, Taylor & Francis Group, LLC, an Informa business, 711 Third Avenue, New York, NY, 10017, USA, and 3 Park Square, Milton Park, Abingdon, OX14 4RN, UK.

Printed in Great Britain

15 14 13 12 11 10 9 8 7 6 5 4 3 2 1

Garland Science
Taylor & Francis Group

Visit our web site at http://www.garlandscience.com

Preface

Plant Pathology is designed to introduce an important and potentially fascinating subject area to a new generation of plant scientists as well as those studying a range of other disciplines. It aligns classic studies and our knowledge to date in the field of plant pathology with the current state of research, all presented in a very accessible format. Topics covered include pathogen taxonomy, the infection process, plant responses to infection, epidemiology, identification of important disease organisms, and disease management. This book provides an excellent introduction to this important field for students studying plant science, horticulture, agriculture and crop production, forestry, or microbiology. Although aimed primarily at undergraduates, it will also serve as a useful reference work for those engaged in research in these subject areas. The text is enhanced by a four-chapter compendium of fungal, bacterial, viral, and other diseases, as well as a range of photographs, photomicrographs, tables, illustrations, and maps, to help the reader to build a detailed picture of this exciting and rapidly evolving topic.

The human race is becoming increasingly distanced from issues such as food security and habitat conservation at a time when these are under greater threat than ever before. In technologically advanced countries, where food supplies are taken for granted and supermarkets provide an ever expanding range of foods from around the world, it is all too easy to become complacent. In developing countries that lack such advanced technology, very little is there for the taking and it is the ability to provide today's meal that is the urgent priority, not the subtle nuances of long-term land management.

Any contemporary book on plant pathology faces significant challenges. This complex topic needs to be addressed in a manner that is relevant and interesting to undergraduate students, while at the same time highlighting how important these complex issues are in relation to the future protection of plant species.

The need for graduates with a thorough grounding in plant pathology is greater than ever before. This book attempts to address this need by making plant pathology accessible to students with no prior knowledge of the subject and setting them on the road to understanding. Many of the current texts on plant pathology, most notably that by George Agrios, are excellent resources for any student who wishes to pursue a career as a plant pathologist. However, we believe that stimulating such an interest in the first place entails getting back to basics. This involves, for example, providing an introduction to cell structure and the relationship between form and function, and a grounding in the identification of plant diseases and recognition of the processes that follow infection. We have used real historical and present-day examples in our case studies, and included up-to-date information on highly topical issues, such as the impact of global climate change on pathogenicity.

We hope that *Plant Pathology* encourages readers to become so engaged with the subject that they are empowered to become involved and make a difference. The future of our planet is dependent on the dedication and ingenuity of the generations to come, and their ability to address the complex issues of food security and land management.

ACKNOWLEDGEMENTS

Grateful thanks to Sonia Cook for her exemplary and timely proof reading.

Thanks go to Roy Moate, Glen Harper and Peter Bond at the Plymouth University Electron Microscopy Laboratory, for all their help in preparation, advice and patient assistance in the use of the Scanning Electron Microscope.

Thanks also go to Ben and Paul Rodgers for allowing us to trample all over their crop fields.

Thanks also to all of those people who have allowed us to use their images.

Any many thanks to the Garland Science staff, especially David Borrowdale, for their patience, help and advice throughout.

DEDICATION

Sarah and Stephen Burchett would like to dedicate this book to Sarah's mother, Muriel Cook (October 1915 – June 2015). She never ceased to encourage us and to take an interest in our work, but sadly didn't get to see the finished article. Love and thanks, Mum.

ONLINE RESOURCES

For instructors, all of the figures from the book are available to download in both PowerPoint® and JPEG format. Contact science@garland.com to gain access to the figures.

Contents

Section 1
Aspects of Plant Pathology

Chapter 1: Concepts and Principles

This chapter covers the historical and contemporary issues relating to plant pathogens. First it outlines the development of scientific knowledge in plant pathology, featuring important historical issues such as the Irish potato famine, ergotism, and Dutch elm disease. The chapter then discusses the impact of disease organisms on food security, outlining the major threats to food supply chains posed by these pathogens, and introduces the role of plant biotechnology and genetically modified organisms as modern tools used to enhance the control of disease.

Chapter 2: Characterization and Taxonomy of Plant Pathogens

This chapter introduces the range of plant pathogens and explores the taxonomy of plant disease organisms, with examples of bacteria, fungi, and viruses as well as some of the other less well-known groups. The chapter also illustrates some of the key biological features of each major taxonomic group.

Chapter 3: Infection Processes

This chapter investigates the dynamics of disease infection, including mechanical and chemical processes. Issues such as enzymatic entry, production of microtoxins, important plant growth regulators, and molecular aspects of infection are discussed, as well as the processes of cellular degradation and the development of structures such as fungal haustoria.

Chapter 4: Plant Responses to Pathogens

This chapter examines how plants respond to disease, starting with the ways in which plants have evolved structural defenses against pathogenesis. It then explores how plants respond at the metabolic level, illustrated by changes in protein metabolism and the acquisition of acquired systemic resistance, and the mechanism of action of applied stimulants such as jasmonic acid and liquid seaweed extract. The chapters also looks at developments in molecular biology, such as the cell-to-cell concept, host recognition, and how this knowledge has enabled scientists to develop plant breeding programs that focus on the production of new plant lines that express novel traits which allow plants to resist disease infection.

Chapter 5: Epidemiology

This chapter looks at how climate and crop factors lead to the spread of plant disease, as well as considering the impact that climate change may have on future trends in disease outbreaks. It also explores the impact of increased genetic uniformity in cropping systems on the likelihood of major disease outbreaks, and illustrates how models in epidemiology are used to predict the course of disease spread.

Chapter 1
Concepts and Principles

In 2011 the human population reached 7 billion, and it continues to increase by approximately 73 million people a year. The majority of this increase is occurring in economically emerging countries that are typified by fragile ecosystems and subsistence agriculture. Crops in these regions are vulnerable to a range of biotic and abiotic perturbations that often result in chronic food shortages and, in worst-case scenarios, famine. Added to this vulnerability is the fact that the human population relies on 12 plant species to provide 75% of the world's food supply, with 50% of the calories provided by three main crops, namely rice, wheat, and corn or maize. Any challenge to these main crops may have a major effect on the ability to meet the demand for food supplies for our burgeoning population.

Plant **pathogens** make a significant contribution to reductions in crop quality and yields, and can have a profound detrimental impact on food security, particularly in those parts of the world that are challenged by food poverty. During the last 40 years, countries from the economically developed regions of the world have invested substantial sums of money and research effort in improving crop yield, and are consequently able to escape the worst effects of disease epidemics. These advances in agricultural technology have resulted in increased yields of staple food crops (Figure 1.1), and are the result of plant breeding programs that improve disease resistance, an increase in resource allocation that has led to improved yields, and improvements in agrochemicals. However, humankind now faces new threats to food security because the rate of increase in crop yields has slowed significantly since 2006, from an average increase of 3.6% per year since the mid-1960s to 1.3% per year in 2010. This decline in the rate of yield increase can be explained by a number of factors, such as losses of cropping area and climate change. However, more importantly, the plants and animals that are used by humankind are reaching their physiological maximum with regard to biological enhancement. All of these factors lead to the conclusion that there is a need for increased activity in areas of agricultural research. Indeed the Food and Agricultural Organization of the United Nations (FAO) has called for a global increase in funding of the order of US$ 85 billion.

During the last 30 years there has been a steady and continuing decline in government spending on research and development in agriculture. In particular, successive governments in the UK have undermined the agricultural sector, resulting in the closure of research institutes (for example, Long Ashton Research Station) and education centers (for example, Seale-Hayne College and Wye College) that were dedicated to the training of future farmers and agricultural researchers. This shortsighted funding strategy has

Figure 1.1 Increase in global yield of small grains during the period 1960–2010.

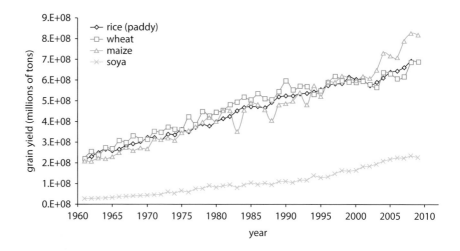

resulted in a loss of skilled people to take agriculture forward in the twenty-first century and beyond. This is a pertinent and important point with regard to plant pathology. Research and training in plant pathology have declined significantly at a time when humankind must increase crop yields by 60–70% by 2050 to meet the growing demand for food supplies.

Plant pathogens continue to challenge crop yields. Losses to plant disease still represent 10–16% of the global harvest, and these losses occur even after the adoption of disease-resistant cultivars and the widespread use of fungicides. This last point illustrates another problem that is often overlooked, namely that plant pathogens respond rapidly to any selection pressure (for example, growing a large area of crops with a single disease-resistant gene, or use of a single active chemical compound in a fungicide) and develop resistance to the armory of weapons at our disposal to protect crops.

There are two other problems facing agriculture and thus food security. First, there is a general lack of understanding by most people of how fragile our food production systems are, and second, decreasing numbers of students are choosing to work in the agricultural sector at all levels. Again this last point is clearly illustrated in developed countries such as the UK and North America, where young school leavers and graduates perceive a career in the agricultural industry as hard work, boring, and unrewarding. It may be hard work, but there are many interesting and rewarding career opportunities in agriculture, ranging from farming to management, consultancy, policy making, and research. With regard to plant pathogens there is now a pressing need to increase activity in primary research and applied science to help to mitigate against any future damaging disease epidemics. In view of this, the aim of this book is to develop knowledge and understanding of plant pathogens.

1.1 Impact of Plant Pathogens on Humans

Historically, humankind has had to contend with plant pathogens since the origins of agriculture around 10,000–12,000 years ago. Indeed **urediniospores** (one of the infectious spore forms) of wheat rust (*Puccinia graminis*) have been dated to 1300 BC from archaeological digs in Israel, and there are biblical accounts of rust epidemics from around 1870 BC. The historical effects of major disease outbreaks on human populations are relatively well documented, but it is difficult to estimate the impact of plant pathogens on human history. Over the last few centuries, one plant pathogen in particular may have changed the course of human history in some unusual and unexpected ways. Ergot, a disease of cereals and other grasses (Research Box 1.1), has caused significant human health problems throughout history.

RESEARCH BOX 1.1 ERGOT: THE "CUCKOO" OF THE PLANT PATHOGEN WORLD

Introduction

Ergots are the **sclerotia** of fungal pathogens of the genus *Claviceps*. Over 40 *Claviceps* species are known, and all of them are parasitic on grasses and rushes, infecting over 400 host species. Few are of economic importance, with notable exceptions, such as *C. africana* and *C. sorghi*, which affect sorghum crops in tropical and subtropical regions, and *C. purpurea*, which is found in temperate regions.

C. purpurea has a long and tainted history, and causes severe illness or even death of humans and animals that inadvertently consume it. Although now well recognized and relatively easy to treat, or at least to avoid, problems associated with ergot infection still arise from time to time. However, as with so many other highly volatile substances, some of its biochemical properties can be exploited and used medicinally, and important medical research is currently ongoing.

Historical Context

Historically, major epidemics of a disease that caused a number of extreme symptoms in human populations have been recorded. This disease manifested itself as two generalized sets of conditions. Gangrenous ergotism caused tissue death, often resulting in sloughed-off flesh, and loss of fingers, toes, or even whole limbs (gangrene), known in the Middle Ages as St Anthony's Fire because of the burning sensations experienced by sufferers. Convulsive ergotism caused convulsions, severe hallucinations, "madness," and death. The latter symptoms were perceived as the work of the devil, demons, or notably of witchcraft, and were almost certainly responsible for the witch trials in Salem, Massachusetts in 1692. These historical records have been associated with ergotism (a disease caused by ingestion of ergot). Each major outbreak can be linked, on the basis of soil horizon records and tree ring records, to periods of cool humid weather conditions when ergot propagation would be particularly successful. Numerous other accounts of convulsive ergotism and gangrenous ergotism have been recorded, including the loss of 20,000 soldiers of the Russian Tsar Peter the Great in the early eighteenth century. It can only be postulated whether the course of global history might have taken a number of different turns were it not for this fungal pathogen. It was not until the late nineteenth century that the causative agent of this disease was first identified.

Life Cycle

C. purpurea infects a number of different grass species, especially cereals, but is particularly prevalent on rye (*Secale cereale*). Although it affects many grass and cereal species, such as wheat and barley, many of these are inbreeders, whereas rye is an outbreeder. This difference in host life cycle aligns with the life cycle of the pathogen, making the rye particularly vulnerable (Figure 1).

The sclerotia lie dormant in the soil over winter, and require conditions of near freezing to below freezing temperatures for a period of about 25 days in order to trigger the production of lipase enzymes that will mobilize the lipids which represent around 50% of the sclerotium tissue. The sclerotia subsequently germinate, favoring temperatures in the range 9–15°C (but inhibited above 18°C), and form perithecial stromal bodies containing asci. Ascospores develop from these asci and are discharged at a time that coincides with anthesis in the rye host. The ascospore mycelia grow specifically towards the vascular tissue of the rachilla (the floret stalk). The duration of flower opening is therefore directly correlated with infection by *C. purpurea*, and this explains why other cereal types, which only open their flowers for short periods of time, are less vulnerable. The proximity to the host vascular system ensures that photosynthate is readily available to the pathogen. At this stage, growth is in an upward direction, and conidial stroma heads (up to 60 from a single ergot) develop beneath the ovary. Within 1 week, unicellular hyaline conidia develop within visible beads of sticky fluid, commonly known as "honeydew," on the florets. The presence of the pathogen up-regulates the movement of water and photosynthate to the infected flowers by creating a sucrose sink and thus increasing the osmotic potential. The honeydew contains a number of different sugars, including glucose, fructose, and sucrose, which attract insects that then act as passive vectors by carrying spores to other host plants. Meanwhile, once in place, the conidial stroma overrides normal development of the ovary into the caryopsis, and the pathogen completely replaces the grain in the ear, although it grows to at least three times the size of the grain—somewhat like a cuckoo in a nest!

Harvest-time

Once the sclerotium is fully developed (Figure 2), the ergot usually falls to the ground, where it begins the life cycle all over again. Sometimes, however, it is harvested along with the crop. In modern times the ergot can be removed by machine, or by floating the grain harvest on saline water, using either sodium chloride or potassium chloride. The grain sinks, but the ergot floats and can then be skimmed off. Significant but highly variable grain yield losses result from *C. purpurea* infection. To avoid perpetuating this problem, farmers can deep-plough, as the ergot will not germinate if it is at a depth of 25 cm or more below the soil surface. Crop rotations using non-grass crops in the second year are also an effective preventive measure, as the ergot will not survive longer than one winter before it needs to germinate. Fungicides can also be effective on some *Claviceps* species. Furthermore, there is some varietal resistance in many crop plants. However, in typical **boom and bust** scenarios, a number of pathogen races are known to occur.

Ergotism

Ergotism is a serious disease caused by ingestion of ergot that is harvested along with the crop. Just 0.03% ergot within the grain is considered to be the limit of acceptability for human consumption, although this strict limit may not necessarily be adhered to for livestock consumption. Low-level ingestion in stock animals can cause symptoms such as loss of appetite (and thus inadequate weight gain), reduced fertility, and poor milk production. At higher concentrations, symptoms vary depending on the levels of different harmful alkaloids present in the ergot. Gangrenous ergotism causes severe cytotoxic damage, often resulting in loss of body extremities

RESEARCH BOX 1.1 ERGOT: THE "CUCKOO" OF THE PLANT PATHOGEN WORLD

Figure 1 Schematic diagram of the life cycle of *Claviceps purpurea*.

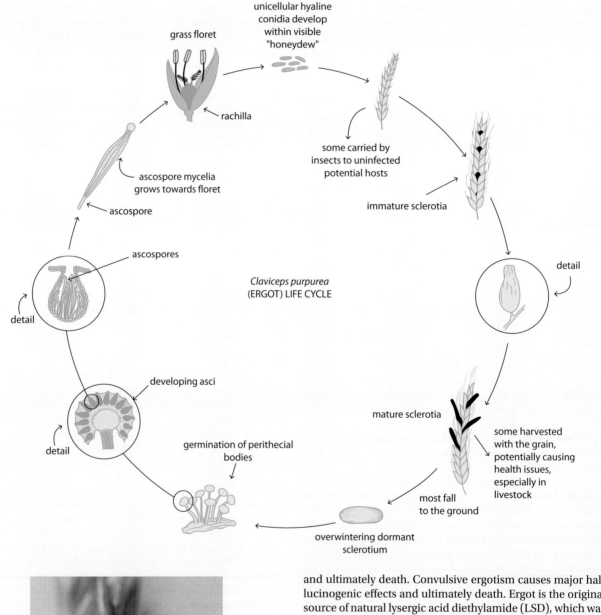

and ultimately death. Convulsive ergotism causes major hallucinogenic effects and ultimately death. Ergot is the original source of natural lysergic acid diethylamide (LSD), which was discovered by Albert Hofmann, and has been used extensively as an illegal drug.

Medicinal Uses

These same properties have enabled pharmacologists to use ergot, in minute and controlled quantities, to produce a number of useful medicines. Around 20,000 kg of ergot-based alkaloids are produced globally each year, 50% of which are harvested from the field. These alkaloids have a tetracyclic structure in the form of an ergoline ring made up of tryptophan, mevalonic acid, and various methyl groups from methionine, and they function by mimicking neurotransmitters such as noradrenaline, dopamine, and serotonin. Two such alkaloids have proved to be especially useful, namely

Figure 2 Fully developed sclerotium of *Claviceps purpurea*.

RESEARCH BOX 1.1 ERGOT: THE "CUCKOO" OF THE PLANT PATHOGEN WORLD

ergometrine, which causes constriction of smooth muscle and is used in particular to stimulate uterine contractions, and ergotamine, which is used mainly as a vasoconstrictor. For many years the properties of these alkaloids have proved effective for treating migraine. More recently they have been used in the treatment of Parkinson's disease, as they function as prolactin inhibitors and enhance natural dopamine production which would otherwise be deficient in patients with this illness. Unfortunately, this treatment has side effects, as it can cause valvular heart disease. Currently research is in progress to manufacture synthetic alternatives, using the natural ergot derivatives as a template.

More recently, well-known major epidemics have included the Irish potato famine and the outbreak of Dutch elm disease in the UK. Both of these diseases caused significant social upheaval.

The first and most well-known example, the Irish potato famine, was caused by the fungus-like pathogen *Phytophthora infestans* (commonly known as late blight), which swept across Ireland in the 1840s. This plant disease caused the deaths of over 1 million people, and compelled a further 2 million to emigrate, mostly to the USA. The second example, Dutch elm disease, did not directly affect food supply but did result in a profound change to the rural landscape of the UK. During the 1960s and 1970s, an outbreak of Dutch elm disease (caused by *Ceratocystis ulmi*, a fungal pathogen spread by the bark beetle, *Scolytus scolytus*) resulted in the widespread loss of English elm (*Ulmus procera*). It was estimated that around 25 million elm trees were lost. This had a significant impact on the rural landscape, removing a rural icon from many regions of the UK. A similar problem, known as sudden oak death (caused by *Phytophthora ramorum*), is now facing managers of stately homes, rural estates, popular public gardens, and woodlands. This problem is not restricted to the UK, as *P. ramorum* has also had an impact on landscapes in the USA and further afield, such as Big Sur in California, where many tanoaks (*Lithocarpus densiflorus*) have succumbed to the disease. Furthermore, *P. ramorum* is not restricted to oak species, as the common name of the disease implies, but is affecting a number of other tree genera, such as the larch (*Larix* species).

These well-known examples illustrate how plant pathogens have both direct and indirect effects on human societies. Another excellent historical example is the way that coffee leaf rust resulted in the UK becoming a nation of tea drinkers. Globally, coffee (*Coffea arabica*) is a highly valuable crop; the global value of green coffee in 2013 was estimated to be US$ 9.8 billion. The major fungal pathogen of coffee is coffee leaf rust (*Hemileia vastatrix*), which leads to total defoliation of coffee plants in the second year of infection. Coffee leaf rust can be found in most coffee-growing regions of the world, including Asia, Central and Latin America, South America, and Africa, including West Africa, from where the disease jumped to Brazil in the 1970s. This westward jump is documented to have occurred between the months of January and April in the late 1960s, as a result of windblown urediniospores. However, the major historical impact of coffee rust epidemics on human society occurred in Ceylon (now Sri Lanka) in 1869. At that time Ceylon was the key coffee producer for the British Empire, and the British were fashionable coffee drinkers. Initial outbreaks of the disease were ignored as the early impacts on plant physiology were minimal, but successive infections led to complete defoliation of the coffee plant and hence crop failure. In 1879, Harry Marshall Ward identified the causative agent and documented the characteristics of the disease. However, this was too late for the coffee farmers in Ceylon and other countries of South-East Asia. As a result, the British promoted tea, this crop eventually replaced the coffee industry in Ceylon, and the British subsequently became a nation of tea drinkers.

Figure 1.2 Global production area of small grains, expressed in millions of hectares, shown in 10-year intervals for the period 1961–2011.

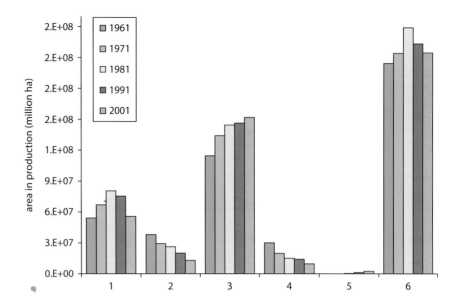

These examples of how plant pathogens have shaped human societies are not solely confined to the history books. Contemporary issues facing humankind as a result of plant pathogens can be seen for a number of our staple crops. Globally, there are six important small grain crops (Figure 1.2), and two of these crops form the bulk of the world's staple foods, namely rice (*Oryza* species) and wheat (*Triticum* species), both of which are susceptible to a range of plant pathogens. Wheat in particular is susceptible to a range of rust pathogens, namely stem or black rust (*Puccinia graminis* Pers. f. sp. *tritici*), stripe or yellow rust (*Puccinia striiformis* Westend. f. sp. *tritici*), and leaf or brown rust (*Puccinia recondita* Rob. ex Desm. f. sp. *tritici*).

Currently, a virulent strain of stem rust, UG99, is spreading across Africa and the Middle East and is threatening Europe. UG99 is known as Ugandan stem rust, as it was first detected in late 1999 in Uganda, from where it started its windborne journey to the Middle East and South Africa (Figure 1.3). This

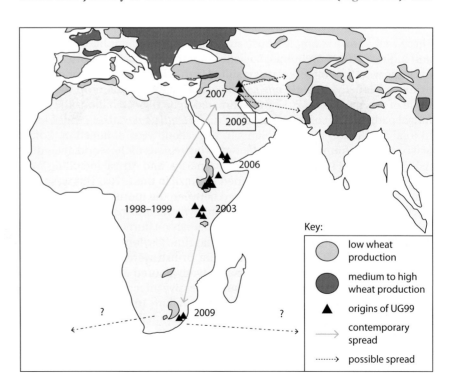

Figure 1.3 Spread of UG99 from East Africa to Southern Europe, the Middle East, and South-East Asia.

outbreak of UG99 is causing significant concerns about the security of future wheat yields, and the estimated potential loss is around 60 million tons—a not insignificant figure. However, more importantly for the people living in the infected regions, there is an immediate impact on food security. Many of the countries that are currently affected by UG99 have a high proportion of subsistence farmers and communities, and it is these individuals and communities who are most vulnerable to such outbreaks.

The impact of rice blast (*Magnaporthe oryzae*) and other plant pathogens in 2003 gave rise to a global yield loss of around 12.2%, and when this is added to the impact of other biotic factors—that is, the effects of pests (15.1%) and weeds (10.2%)—the combined total yield loss for 2003 is calculated to be 37.5%. These figures illustrate how important it is to maintain and enhance our research into plant pathogens.

1.2 Plant Pathology and Pathogens

Plant pathology is a branch of science concerned with the study of the organisms that cause disease, the **epidemiology** of plant pathogens, and the control measures that can be adopted to alleviate the effects of plant disease. However, before we continue further it is helpful to define what we mean by plant pathology in terms of the impacts on plants and the causative agents involved.

Plant disease is usually defined as being caused by any organism that alters the normal physiological process in plants, which may lead to loss of leaf surface area, loss of yield components (such as grains and fruits), necrosis of sections of the plant or total plant death. Pathogens are the main causes of disease in plants, although pathogenicity is a complex topic that involves a number of other contributory factors, including ambient environmental conditions and host susceptibility, and it will be described in some detail in this text. Many plant pathogens give rise to an array of external symptoms that range from chlorosis or yellowing of the leaves to water-soaked and necrotic tissue, and mats of white mycelia (Figure 1.4). Rust pathogens result in regions of leaf and stem tissue expressing rows or patches of brown, yellow,

Figure 1.4 Morphology of various infectious agents on plants. (a) Gray mold (*Botrytis cinerea*) on fruit of raspberry (*Rubus* species); (b) brown rot (*Monilinia* species) on fruit of apple (*Malus* species); (c) mildew on leaf of greater plantain (*Plantago major*); (d) leaf spot (*Mycosphaerella fragariae*) on leaves of strawberry (*Fragaria* species); (e) black rot (*Xanthamonas* species) on brassica leaf. A color version of this figure can be found in the color plate section at the end of the book.

(a)

(b)

(c)

(d)

(e)

Figure 1.5 Morphological effects of virus infection in plants. (a) Viral distortion on leaf of lettuce (*Lactuca sativa*); (b) color-break virus in flower of tulip (*Tulipa* species). Both effects are caused by potyviruses.

orange, and black pustules (Figure 1.4). Other pathogens, generally viruses, result in distortion of leaves, stems, and flowers (Figure 1.5). However, some external symptoms can be confused with signs of nutrient deficiency, such as the expression of chlorosis seen in *Camellia* species (Figure 1.6), which in many cases is the result of iron or magnesium deficiency. The ability to recognize these symptoms and correlate them with known pathogens is a key skill that all plant pathologists have to develop, and this theme is further developed in the compendium chapters of this book. However, in order to acquire these skills it is essential that students understand the range of organisms that cause disease in plants.

The above description of plant disease is quite wide-reaching, and could encompass a whole range of different agents, including fungi, bacteria, viruses, nematodes, insects, and abiotic factors. For the purposes of this text the term "plant pathology" will be restricted to the fungi and fungus-like organisms, the bacteria, the viruses, and other parasitic organisms (Table 1.1). Plant pests have had to be excluded because they are an extensive topic that is beyond the scope of this volume. Nevertheless, a brief review of what is meant by the term "plant pest" is warranted, as a number of plant pathogens can be spread by pest organisms, which are known as **vectors**.

A plant pest can be generally defined as any organism that has a detrimental impact on plant and crop health. These organisms are dominated by the invertebrates and, more specifically in the tropics, by the arthropods (including insects and mites). However, other organisms are problematic for many growers. These include mammals such as the European rabbit (*Oryctolagus*

Figure 1.6 (a) Healthy *Camellia* leaf; (b) *Camellia* leaf showing magnesium deficiency. Loss of chlorophyll can be attributed to a number of different conditions other than disease, including mineral deficiencies.

Table 1.1 Common plant pathogens by taxa.

	Pathogen	Common Name	Region
Fungi	*Alternaria alternata*	Stem canker of tomato	Wordwide
	Ceratocystis ulmi	Dutch elm disease	Northern hemisphere
	Claviceps purpurea	Ergot	Worldwide
	Fusarium oxysporum	Cotton wilt	Wordwide
	Gaeumannomyces graminis	Take-all	Worldwide
	Pseudocercosporella herpotrichoides	Eyespot	Worldwide
	Puccinia antirrhini	Rust of antirrhinum (snapdragon)	Wordwide
	Puccinia coronata	Crown rust of oats	Worldwide
	Puccinia striiformis	Stripe rust of wheat	Wordwide
	Pyrenophora tritici-repentis	Wheat tan spot	North America
	Rhizopus species	Soft rot of fruit	Worldwide
	Sclerotinia sclerotiorum	Stem rot	Worldwide
	Verticillium dahlia	Vascular wilt of potatoes and strawberries	Wordwide
Bacteria	*Agrobacterium tumefaciens*	Crown gall disease	Worldwide
	Colletotrichum gloeosporioides	Bitter rot of apple	Worldwide
	Corynebacterium flaccumfaciens	Bacterial wilt of bean	Widespread
	Corynebacterium michiganense	Bacterial canker of tomato	Worldwide
	Corynebacterium sepedonicum	Bacterial ring rot of potato	North America and Europe
	Erwinia amylovora	Fire blight	Worldwide
	Erwinia dissolvens	Corn rot	North America, Canada, Europe
	Erwinia stewartii	Wilt of maize	Worldwide
	Erwinia tracheiphila	Cucumber wilt	North America, Europe, South Africa, and Japan
	Pseudomonas alboprecipitans	Bacterial spot on cereals	North America, Asia
	Pseudomonas solanacearum	"Moko" disease of bananas	Tropical western hemisphere
	Pseudomonas syringae pathovar (pv.) *pisi*	Bacterial blight of pea	Worldwide
	Xanthomonas campestris	Black rot	Worldwide
	Xanthomonas carotae	Bacterial blight of carrot	Worldwide
	Xanthomonas citri	Citrus canker	Worldwide
	Xanthomonas arboricola pv. *pruni*	Canker of stone fruits	Most stone-fruit-growing regions
Virus		Apple stem grooving virus	Worldwide
		Apple stem pitting	Worldwide
		Barley stripe mosaic virus	Worldwide
		Barley yellow dwarf virus	Worldwide
		Bean common mosaic virus	Worldwide
		Cabbage ring necrosis	Worldwide
		Citrus tristeza virus	All citrus-growing regions
		Grapevine fanleaf virus	North America and Europe
		Lettuce infectious yellow virus	South West America and Mexico
		Lettuce mosaic virus	Worldwide
		Maize dwarf mosaic virus	Worldwide
		Onion yellow dwarf virus	Worldwide
		Plum pox virus	Worldwide
		Potato mop-top virus	Worldwide

(Continued)

Table 1.1 (*Continued*) **Common plant pathogens by taxa.**

	Pathogen	Common Name	Region
		Potato virus Y	Worldwide
		Sugarcane mosaic virus	Worldwide
		Tobacco mosaic virus	Worldwide
		Tomato mosaic virus	Worldwide
		Tomato spotted wilt virus	Worldwide
Other	*Peronospora destructor*	Downy mildew	Worldwide
	Phytophthora infestans	Potato blight	Worldwide
	Plasmodiophora brassicae	Club root	Worldwide
	Spongospora subterranea	Potato wart disease	Worldwide

cuniculus), which can lead to significant losses of crops adjacent to field boundaries (Figure 1.7), as well as mice and other small mammals, which cause significant losses of stored grains, and in Africa migrating elephants (*Loxodonta* species), which can be very problematic where field crops are grown. Another iconic animal that is regarded as a pest is the orang-utan (*Pongo pygmaeus* and *P. abelii*). Sadly for this animal, the rapid expansion of oil palm (*Elaeis guineensis*) plantations has not only encroached on the habitat of orang-utan, but has also led to human–animal conflict between oil palm plantation managers and orang-utans, because these animals wander into plantations in search of food. In wetter regions, molluscs such as slugs and snails cause significant damage to young establishing crops, and are the bane of many ornamental gardeners' lives.

However, it is the insect pests that cause the most significant damage to crops worldwide, and particularly in the tropics. With regard to the transmission of plant pathogens it is the leafhoppers and aphids that will be highlighted later on in this text.

Plant pathogens are generally microscopic or sub-microscopic organisms that infect a range of plant species, leading to partial or complete plant death. Economically these pathogens are truly problematic; in 2003 the global, pre-harvest loss of yield exceeded US$ 220 billion. These organisms can infect all plant organs and tissues, including roots, stems, leaves, flowers, fruit, and seed (Figure 1.8). They also infect internal tissues such as the

Figure 1.7 Typical field-scale rabbit damage to oilseed rape (*Brassica napus*) crop.

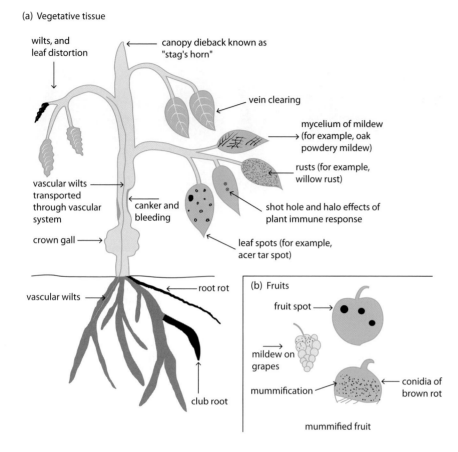

(a) Vegetative tissue

wilts, and leaf distortion

canopy dieback known as "stag's horn"

vein clearing

mycelium of mildew (for example, oak powdery mildew)

rusts (for example, willow rust)

vascular wilts transported through vascular system

canker and bleeding

shot hole and halo effects of plant immune response

crown gall

leaf spots (for example, acer tar spot)

vascular wilts

root rot

club root

(b) Fruits

fruit spot

mildew on grapes

mummification

conidia of brown rot

mummified fruit

Figure 1.8 Schematic diagram of infected plant organs showing a range of potential disease threats.

vascular system and the mesophyll tissue, including the stomata and the stomatal chambers; these plant structures are the main routes of entry for a number of pathogens. The symptoms expressed by the diseased plant are dependent on the type of parasitic relationship that exists between the plant and the pathogen. These can be summarized as follows:

- **Biotrophs**: these are parasites that grow and reproduce in nature only on living hosts, and are known as obligate parasites. They include the **fastidious** bacteria, viruses, and fungal pathogens, such as some of the rusts and mildews.

- **Hemibiotrophs**: some biotrophs can subsist as saprophytes on dead material for part of their life cycle, and are known as non-obligate parasites (or facultative saprophytes). They include some of the rusts and mildews.

- **Necrotrophs**: these pathogens kill the host cell and then feed off the cellular remains. They are known as facultative parasites, and include *Alternaria* and *Erwinia* species.

Each of these parasitic relationships results in a range of diagnostic symptoms (**Figure 1.9** and **Figure 1.10**) that can aid plant pathologists in identifying the causative agent.

Plant life cycles can be categorized into three groups—annual, biennial, and perennial. Crops are grown from all three of these groups, and are challenged by a huge array of plant pathogens (**Table 1.2**).

In order to appreciate the complex nature of plant disease and the impact of these organisms on plants and crops, it is helpful to define the normal physiological functions of a healthy plant and scale these up to a cropping

1.1 The Father of Microbiology

Antonie van Leeuwenhoek (1632–1723) was a Dutch textile trader. For his business he used glass pearls to closely observe the quality of his cloth. Influenced both by the depth of these observations and by Robert Hooke's drawings in *Micrographia* he developed high-quality, high-resolution lenses, from which he built his own microscopes. Fascinated by Hooke's work, he reproduced a number of his drawn images from his own observations. He then went on to observe many unicellular organisms, including protists in 1674, and he was the first person to record bacteria (which he called "animalcules"), in 1676, as well as many microscopic multicellular organisms, insect compound eyes, red blood cells, and spermatozoa, to name just a few. He never revealed how he developed his lenses, so his method died with him, but they were of such high quality that it would be another 100 years before technology caught up with him and bacteria were observed again.

Figure 1.9 Infection (a) rust on onion leaves; and (b) mildew on wheat leaves. A color version of this figure can be found in the color plate section at the end of the book.

(a)

(b)

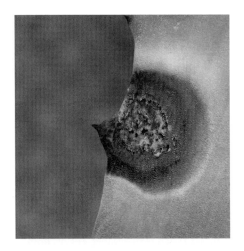

Figure 1.10 Typical *Alternaria* infection on *Agave* species leaf; note the concentric rings.

Figure 1.11 Germinating seedlings of wheat (*Triticum* species).

system. In this case the crop model will be based on winter wheat under UK growing conditions.

Healthy plants of all types progress through the following defined growth stages in order to complete their life cycle: (1) seedling and establishment phase; (2) vegetative phase; (3) flower development phase; (4) reproductive phase; (5) pollination and seed set.

In order for wheat plants to complete their life cycle, they must first germinate and establish their root and shoot systems (Figure 1.11). This is a vulnerable stage for many plants, as they are very susceptible to **damping-off** diseases caused by pathogens such as *Pythium* species and *Rhizoctonia solani*. Plants are more vulnerable during this time because the young growing tissues have not fully hardened, and natural defenses such as mature cuticular wax or hairs have not yet developed. The above-mentioned disease organisms are soilborne pathogens that can invade both roots and stems of young seedlings, causing plant death (expressed in young seedlings as darkened water-soaked tissue, stem lesions, and seedling dieback in more or less circular patches), and hence the young plants or crop fail to establish. Assuming that the young seedling escapes the ravages of damping-off, it then goes through a vegetative growth stage, which for winter wheat has been fully characterized by crop scientists (Figure 1.12). There are several disease organisms that challenge the vegetative growth stage of plants (Table 1.2), and for cereal crops such as winter wheat these include many biotrophic pathogens that interfere with photosynthetic processes.

During the vegetative growth stage the plant must maximize solar energy gain in order to achieve maximum photosynthetic efficiency. Through the processes of evolution, extant plants have developed unique canopy architectures and photosynthetic mechanisms that enhance their photosynthetic efficiency. In winter wheat the leaves of young plants form a rosette and present their leaves perpendicular to the sun's rays. This canopy architecture enables young plants to capture light energy from the weak winter sun with minimal competition for space by neighboring leaves. During winter the young wheat plants also produce side shoots known as **tillers** that further enhance light harvesting. As the young plant matures, it extends its stem (in the UK this occurs from April to May) and simultaneously it sacrifices a number of tillers (winter wheat can produce up to eight tillers, but at the end of stem extension agronomists aim for two tillers) to reduce competition for light. Finally, in the vegetative growth stage the now maturing wheat plant produces its final leaf, known as the flag leaf. This is the

Table 1.2 Examples of crop diseases at different plant growth phases.

	Growth Phase	Scientific Name	Disease (Common Name)
Wheat	Seedling	*Erysiphe graminis*	Powdery mildew
		Puccinia graminis	Black stem rust
		Puccinia recondite	Brown leaf rust
		Puccinia striiformis	Yellow or stripe rust
		Pythium species	Root rot
	Vegetative	*Cochliobolus sativus,* conidial stage (*Helminthosporium sativum*)	Spot blotch or foot rot
		Erysiphe graminis	Powdery mildew
		Fusarium culmorum and *F. avenaceum*	Seedling blight
		Gaeumannomyces graminis	Take-all
		Pseudocercosporella herpotrichoides	Eyespot
		Septoria nodorum	Glume blotch
	Stem elongation	*Gaeumannomyces graminis*	Take-all
		Pseudocercosporella herpotrichoides	Eyespot
		Puccinia graminis	Black stem rust
		Puccinia recondite	Brown leaf rust
		Septoria nodorum	Glume blotch
	Booting, flowering, and grain fill	*Cladosporium* species	Sooty mold
		Claviceps purpurea	Ergots
		Tilletia caries and other *Tilletia* species	Bunts
		Ustilago nuda	Smuts
		Xanthomonas campestris pv. *Translucens*	Black chaff
Cauliflower	Seedling	*Rhizoctonia* species	Bottom rot
		Phytophthora species	Root rot
		Plasmodiophora brassicae	Club root
			Cauliflower mosaic virus
	Vegetative		Cauliflower mosaic virus
	Flowering	*Pseudomonas syringae*	Leaf spot
		Peronospora parasitica	Downy mildew
		Various	Mold
		Various	Canker
		Various	White rust
		Verticillium species	Wilt
			Cauliflower mosaic virus
Strawberry	Seedling	*Phytophthora fragariae*	Strawberry red stele disease

largest leaf, and it tends to be held perpendicular to the stem axis. The role of the flag leaf is to maximize photosynthesis, as 85% of the photosynthate produced by the flag leaf is used in the development of the grain. At every stage of vegetative development the wheat plant is challenged by a number of biotrophic pathogens, but most importantly by the rusts and mildews that can directly interfere with photosynthetic processes. Both of these pathogens cause a significant reduction in healthy green leaf area. The mildews develop an extensive mycelium, which can completely cover the leaf surfaces and thus reduces the light-harvesting area. Infection with rusts

Figure 1.12 Schematic diagram of the key stages in cereal growth, based on the decimal growth stage (GS).

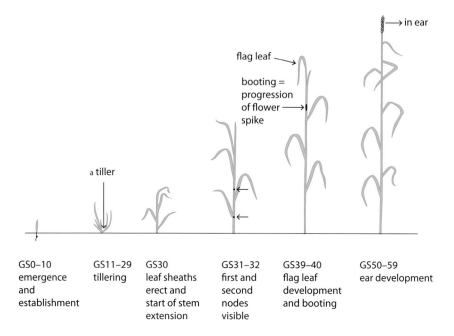

| GS0–10 emergence and establishment | GS11–29 tillering | GS30 leaf sheaths erect and start of stem extension | GS31–32 first and second nodes visible | GS39–40 flag leaf development and booting | GS50–59 ear development |

can lead to large lesions and patches of germinating rust pustules (Figure 1.13) that again reduce the green leaf area. However, these pathogens also redirect the host's photosynthate for their own use, and thus reduce photosynthetic efficiency by harvesting the host's carbon energy. With regard to the flag leaf crop, agronomists aim to keep this leaf free from all disease, and they therefore target disease control throughout the vegetative growth phase of the crop.

Other important fungal pathogens during the vegetative growth phase of wheat include eyespot disease (*Pseudocercosporella herpotrichoides*), which infects the lower section of the stem and thus reduces the efficiency of water uptake and weakens the standing power of the plant, leading to **lodging** just before harvest. With regard to management of these pathogenic problems during the vegetative growth phase, agronomists aim to reduce the carry-over inoculum of disease agents by targeting specific growth stages of the crop and pathogens (see Chapter 10).

Vegetative damage to plants and crops is not only caused by fungal pathogens. Many plants are damaged by bacteria and viruses, and both bacterial and viral diseases cause a significant reduction in green leaf area. For example, barley yellow dwarf virus results in yellowing of leaves and stunted plants. Bacterial diseases often cause leaf spotting and chlorotic and necrotic lesions on leaves. For example, many brassica crops are affected by the bacterial pathogen *Xanthomonas campestris* that causes brassica black rot. This bacterial pathogen infects a wide range of plant species, many of which are of economic importance.

Another important group of pathogens that cause significant damage to plants during the vegetative growth phase consists of the vascular wilts (*Fusarium* species) and other root and vascular diseases, such as take-all (*Gaeumannomyces graminis*) of wheat. If these organisms infect the plant before it has completed the vegetative growth phase they can significantly reduce its photosynthetic efficiency by reducing water uptake, and also restrict plant growth due to loss of root area.

The final stage in the wheat life cycle consists of flowering, pollination, and grain filling, and for the grower this is a key economic stage that has

Figure 1.13 Rust pustules on leaves of wheat (*Triticum* species). A color version of this figure can be found in the color plate section at the end of the book.

to be managed carefully, as a number of pathogens can infect the grain, such as ergots, the smuts, bunts, and glume blotch. These pathogens will cause reductions in grain yield and final grain quality. For example, glume blotch (*Septoria* species) invades the flower spike late in the life cycle of the crop, producing resting spores known as pycnidia, which eventually produce **conidia** that can then reinfect a new crop. In the UK, *S. tritici* is the key pathogen of wheat crops grown in the south-west region of England. Another problematic disease of cereal grains is loose smut (*Ustilago nuda* and *U. tritici*). The mycelium of *Ustilago* species keeps pace with the growing point of the cereal plant, and eventually the mycelium invades the flower spike and subsequently the young kernels of the ear. As the kernels mature, the mycelium of *Ustilago* develops reproductive spores known as teliospores, which eventually fill the kernels of the ear, and in time the membrane that anchors the kernels to the **rachis** breaks, and wind blows the maturing teliospores onto uninfected cereal grains. Subsequently the teliospores that have infected fresh grains produce new mycelium that invades the embryo, thereby perpetuating inoculum carry-over to the next generation.

There are many examples of pathogens infecting flowers, and in ornamental plants such as rose and clematis the biotrophic mildews can cause unsightly infections that reduce the aesthetic quality (Figure 6.4d) of the blooms and thus final value of the crop for growers.

1.3 Summary

Plant pathology is a complex multidisciplinary subject that requires a knowledge of the agricultural and horticultural sector, the plant species grown, and their associated production systems. The plant pathologist then needs to gain an understanding of the range of pathogens and associated vectors, and the symptoms that these pathogens induce in infected plants. An understanding of disease progression within a crop and ultimately within a cropping region is also required. As we have seen in this chapter, pathogens can cause significant loss of crops, and continue to do so. We have endeavored to set the scene for this thought-provoking subject area by introducing some key and fundamental concepts that underpin plant pathology, including a historical perspective of plant pathogens and the impact that these organisms have had on human societies. The concept of the model crop plant, in this case wheat, and the ways in which pathogens can interfere with every stage in the life cycle of the crop were then explored. Parallels were drawn with other common plant production systems and with natural plant communities in order to highlight the widespread nature of plant disease.

1.4 What You Will Learn in This Book

The starting point for any student of plant pathology is to develop a knowledge of the taxonomy of plant pathogens (Chapter 2), and then to gain a deeper knowledge and understanding of plant physiology and subsequent perturbations in plant health caused by disease (Chapter 3). Once this knowledge has been acquired, the next step is to develop an understanding of how plants defend themselves against disease (Chapter 4), and from this point we can start to exploit this knowledge by forecasting disease outbreaks (Chapter 5). These disease models, coupled with an understanding of the common causative agents (Chapters 6, 7, 8, and 9, which form the compendium), can then provide information about how we can attempt to control disease (Chapters 10 and 11). Finally, we can reflect on and evaluate the future trends in agriculture and plant pathology (Chapter 12) that will help to reinforce this discipline for future generations.

Further Reading

Agrios GN (2005) Plant Pathology, 5th ed. Elsevier Academic Press.

Anon. (2006) A Self-Guided Tour of Sudden Oak Death (*Phytophthora ramorum*) Along the Big Sur Coast, Monterey County, CA. www.suddenoakdeath.org/pdf/3.06.Big%20Sur%20self-guided%20field%20trip.pdf

Bushnell WR & Roelfs AP (eds) (1984) The Cereal Rusts. Volume 1: Origins, Specificity, Structure, and Physiology. Academic Press.

Chakraborty S & Newton AC (2011) Climate change, plant diseases and food security: an overview. *Plant Pathol* 60:2–14.

Famous Scientists. Antonie van Leeuwenhoek. www.famousscientists.org/antonie-van-leeuwenhoek/

Food and Agriculture Organization of the United Nations (2008) Climate Change: Implications for Food Safety. ftp.fao.org/docrep/fao/010/i0195e/i0195e00.pdf

Food and Agriculture Organization of the United Nations (2014) AGP – FAO Wheat Rust Disease Global Programme. www.fao.org/agriculture/crops/core-themes/theme/pests/wrdgp/en/

Food and Agriculture Organization of the United Nations (2017) Value of Agricultural Production. Global Value of Green Coffee. www.fao.org/faostat/en/#data/QV

Hulvova H, Galuszka P, Frebortova J & Frebort I (2012) Parasitic fungus *Claviceps* as a source for biotechnological production of ergot alkaloids. *Biotechnol Adv* 31:79–89.

Jones DG (1987) Plant Pathology: Principles and Practice. Open University Press.

Jones DG & Clifford BC (1978) Cereal Diseases: Their Pathology and Control. BASF Agrochemical Division.

Kim JY, Chung EJ, Park SW & Lee WY (2006) Valvular heart disease in Parkinson's disease treated with ergot derivative dopamine agonists. *Mov Disord* 8:1261–1264.

McIntosh RA, Wellings CR & Park RF (1995). Wheat Rusts: An Atlas of Resistance Genes. CSIRO. www.globalrust.org/sites/default/files/wheat_rust_atlas_full.pdf

Roelfs AP and Bushnell WR (eds) (1985) The Cereal Rusts. Volume 2: Diseases, Distribution, Epidemiology, and Control. Academic Press.

Strange NS & Scott PR (2005) Plant disease: a threat to global food security. *Annu Rev Phytopathol* 43:83–116.

Vermeulen SJ, Aggarwal PK, Ainslie A et al. (2012) Options for support to agriculture and food security under climate change. *Environ Sci Policy* 15:136–144.

Webster J & Weber RWS (2007) Introduction to Fungi, 3rd ed. Cambridge University Press.

Chapter 2

Characterization and Taxonomy of Plant Pathogens

The imperative of all living organisms is survival, both in terms of self-preservation and for the survival of the species. This fact is no less true for pathogenic organisms, but in these cases, survival brings them into conflict with the hosts upon which their survival depends. The survival of such organisms is therefore inexorably bound, and thus damaging, to the life cycle of their hosts, and it then causes pathogenesis, often leading to disease.

Such host–pathogen interactions can be:

- Non-specific—where a pathogen can infect a variety of hosts (for example, the mildews)

- Specific—where a pathogen may infect a specific group of hosts (for example, the cereal rusts that infect only the grasses, including cereals)

- Highly specific—where a pathogen only infects a very narrow range of similar hosts (for example, peach X disease, which is an important disease of peach trees, nectarines, and certain types of cherry).

2.1 Naming the Disease

In order to understand how and why these interactions have come into being, it is important to first comprehend the **taxonomy** of pathogenic organisms. Taxonomy is the study of biological classification (the hierarchy of life), and this classification is derived in a number of ways, including ecological niche, morphological uniqueness, and the genetic makeup of species in terms of their relatedness to one another. To complicate the issue, entirely different species that are incapable of interbreeding can be morphologically identical. These are known as **cryptic species.** As species evolve, the classification groupings become ever wider. Furthermore, genetic research has improved the accuracy with which these taxonomic ranks can be defined, so what was in Darwin's time simply based on observations of similarities, and was by definition highly subjective, has now become a highly complex scientific discipline based mainly on **DNA**, **RNA**, and/or protein analyses.

The **Five Kingdom Classification System**, consisting of the kingdoms Animalia (animals), Plantae (plants), and Mycota (fungi), together with the **Protista** and the Monera, was until recent years the most commonly used set of taxonomic groupings. However, most biologists now use a system of three

Figure 2.1 Generalized taxonomy of plant pathogens. (Note that the list is not exhaustive.)

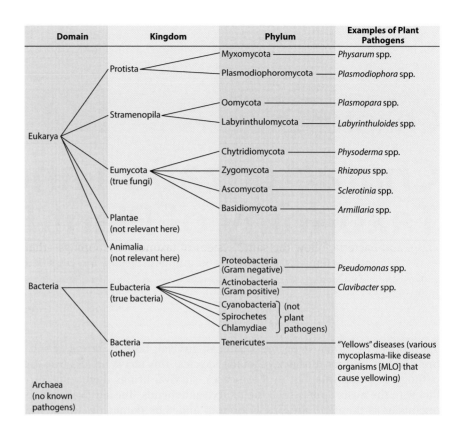

domains, namely the Archaea, the Bacteria, and the Eukarya, followed by a multi-kingdom system, which then precedes the previously established **phylum** (the term division is generally used in place of phylum in plants), class, order, family, genus, and species. (In some texts the term division is also used for the fungi.) Genus and species form the basis of the **binomial system**. These ranks may be further divided into super- or subgroups—for example, a subphylum of the phylum Chordata would be Vertebrata, and so on. Taxonomic groupings are generally illustrated by means of phylogenetic trees (**phylogeny** being the evolutionary history of species), where the branching becomes more defined as the separation of relatedness becomes more defined, and a branching in the phylogenetic tree correlates with the divergence into new species. Figure 2.1 shows a simplified phylogeny that illustrates this concept in principle using examples of just a few of the plant pathogens. Students should also be aware that as recently as 2011 a new code, known as the Melbourne Code, was published to assist workers in defining taxonomic groups.

The domain Archaea (once classified as Archaebacteria) consists entirely of **prokaryotes** and, although ubiquitous, has had no pathogenic species identified to date. The domain Bacteria also consists solely of prokaryotes, with many pathogenic representatives. The domain Eukarya consists of all other life forms not included in the other two **domains**, and therefore includes all pathogenic species other than bacteria and viruses. (Viruses sit outside these groupings for reasons that will be discussed later.)

2.2 Relating Name to Form and to Function

What makes the topic of taxonomy so important in relation to plant pathogens is that because the life cycles of pathogens are so inextricably bound to the life cycles of their hosts, the partnerships inevitably undergo coevolution. As the host develops the means of defense against the pathogen, so the pathogen must inevitably develop the means to respond to such defenses, and again the host must attempt to counteract the pathogen's new response, and so on. This concept, known as the **boom and bust** theory, is discussed further in Chapter 4. However, at this point it becomes clear that the more specific the interaction of the pathogen with its host, the more vital to survival is its developmental evolution, and thus the more relevant taxonomic understanding is to the student of plant pathology.

Here we introduce the taxonomy of the following:

- Fungi

- Bacteria

- Viruses

- Other plant pathogens.

Fungi and bacteria follow the same system of taxonomic groupings—that is, the binomial system—as the plants and animals. Viruses are classified in a completely different manner for reasons that will become clear later in this chapter. In addition, there are a number of other relevant organisms that now, in the light of advances in taxonomic understanding, fall outside the "big three" pathogens (that is, fungi, bacteria, and viruses); these will be highlighted here. In addition, the key biological features that define the cited examples of taxonomic groups will be illustrated in context.

2.3 Fungi

Introduction

Well over 100,000 fungal species have been categorized, although estimates indicate that there are potentially millions. Because fungi are so ubiquitous and diverse there is simply no means of accurately enumerating the group. In this section we provide a general overview of fungal taxonomy while also highlighting the key genera associated with plant pathology.

Fungal Structure

It is important for students to have a knowledge of the basic structure of fungi at the cellular level in order to understand the means by which hyphal development and thus growth of the mycelium facilitates infection of the host. This is discussed in greater detail in Chapter 3. However, the differences in the basic cellular structure of fungi provide a good indication of taxonomic groupings. This is clearly indicated in the descriptions of fungal phyla below.

The main component of multicellular fungi is the **mycelium**, which often forms a mat composed of branching filamentous cells called **hyphae**. Figure 2.2 shows the hyphal structures of *Monilinia fructigena* on an apple. These cells are structured from tubular cell walls that surround the plasma membrane and the cytoplasm, and within the cytoplasm there are numerous cell organelles. The hyphal cells of many fungi have a unique capacity to elongate, sometimes up to many meters in length. Cell growth is limited to the cell apex, resulting in elongation rather than an increase in girth. This ensures the largest possible surface area to volume ratio, to maximize absorption from the host tissue. Nutrients that promote this elongation process are passed to the growing tip by **cytoplasmic streaming**, and in the case of most fungi, fortifying cross walls called **septa** form periodically through the hyphae, effectively creating semi-separated cell units (Figure 2.3). These septa either do not completely traverse the cell, or have perforations so that materials can flow with relative ease through the system, although many species, especially those belonging to the Ascomycota, can if necessary plug these with a peroxisome-based microbody called a **Woronin body** to prevent major losses of cell contents to the outside if any damage occurs to the organism. There are also unicellular fungi, known as yeasts.

Figure 2.2 A well-developed mycelial mat of *Monilinia fructigena* (brown rot) on apple.

Figure 2.3 Generalized diagram of a hyphal cell of a pathogenic fungus, showing concentration of vesicles at the tip.

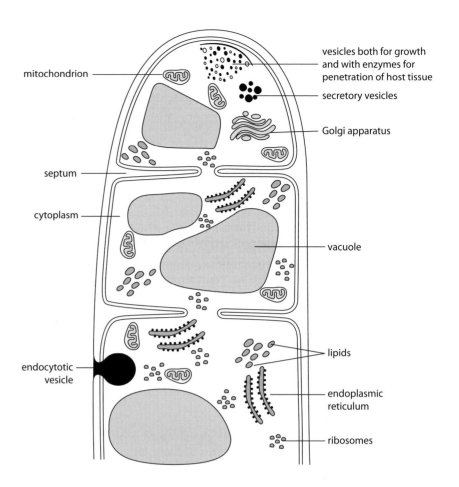

Classification of Fungi

Historically, fungi were thought to be non-photosynthetic plants, an error that is understandable given the morphological similarities and the fact that they produce spores. However, modern diagnostic methods and tools have allowed us to analyze the cell components in greater detail. It has been found that fungal cells share a number of similarities with plant cells—for example, the presence of cellulose as one of the key structural components of the cell wall (discussed in more detail in Chapter 3). However, there are also some similarities with animal cells—for example, one of the other cell wall components is chitin (normally associated with arthropods). In addition, there are components that are unique to the fungi—for example, hydrophobins, which are cysteine-rich proteins that are only found in fungal cells. These self-assembling proteins are involved in the formation of fungal aerial structures, and they also form hydrophobic–hydrophilic interfaces with the surrounding host tissue or with the immediate environment. As a result of these significant differences, the fungi came to stand alone as one of the taxonomic kingdoms.

In addition, there are a number of fungus-like organisms that were once categorized with the true fungi, but which are now known to belong to other kingdoms. Examples of these are the phylum Myxomycota (which includes the slime molds), that now belongs to the kingdom Protista, and the phylum Oomycetes (which includes the water molds and downy mildews), that now belongs to the kingdom **Stramenopiles**. (The fact that some of the fungus-like organisms are now placed in the kingdom Protista is almost a regressive point of interest, as it is widely accepted that the fungi have evolved from ancient aquatic protist ancestors.) These are discussed in greater detail below in the section on other plant pathogens.

Here we shall look specifically at the true fungi—the **Eumycota**. These are subdivided into four phyla:

- Chytridiomycota
- Zygomycota
- Ascomycota
- Basidiomycota.

Of the 100,000 or so fungal species that have been described, more than 90% are **saprophytic** (that is, they rely specifically on decaying organic matter for their nutritional requirements). However, over 8000 species are known to be parasitic on plants (indeed pathogenic fungi are the causative agents of more than 75% of all plant diseases), and these have representatives from each of the four phyla. **Figure 2.4** shows a simplified phylogeny of fungal plant pathogens.

- Chytridiomycota (the chytrids). These are characterized by their flagellated **zoospores** (each zoospore bears a single flagellum), which are unique among the Eumycota, although more common in some of the fungus-like groups discussed later. The chytrids have either rounded or elongated hyphae that do not form septa, and which are therefore described as **coenocytic hyphae**. Some orders also lack true mycelia, and have **hyphal rhizoids** instead. They are typically found in all soil types and in freshwater environments. Four genera in this phylum are known to be pathogenic in plants, namely *Physoderma*, *Urophlyctis*, *Synchytrium*, and *Olpidium* (for example, *O. brassicae*, which causes root rot in cabbages, but in addition often acts as a vector for virus pathogens).

Note: *Synchytrium endobioticum* is a species of the phylum Chytridiomycota that causes potato wart disease. This disease is listed as a potential

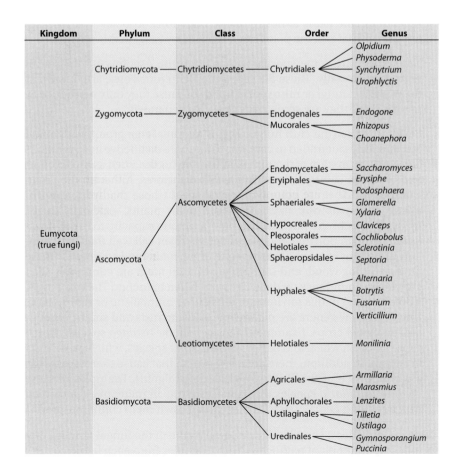

Figure 2.4 Simplified phylogeny of fungal plant pathogens down to the level of genus. The level of family is excluded from this table as familial names largely follow a standard convention (for example, the genus *Olpidium* is in the family Olpidiaceae, and the genus *Erysiphe* is in the family Erysiphaceae). Furthermore, scientific opinions on fungal familial names are notably disparate. (Note that the list is not exhaustive.)

bioterrorism threat, as deliberate large-scale infection could potentially create food security issues in countries where potatoes are a dietary staple.

- Zygomycota (the bread molds). This group is characterized by a non-motile phase in its reproductive cycle. The visible sporangia that are commonly seen on moldy bread and fruit are asexual. These structures disperse spores that develop into mycelia, which later fuse to form sexually reproductive **zygospores** within zygosporangia. These can remain in a very resilient resting phase until environmental conditions are favorable for growth. The resilient nature of this group ensures its survival. One of the main pathogenic genera in this phylum is *Rhizopus*, species of which cause soft rots in fruits and some vegetables.

  (A small subgroup of the Zygomycota has recently been reclassified into a fifth phylum, or is sometimes classified as a subphylum of the Zygomycota. All members of this subgroup, the Glomeromycota, that have been identified so far are arbuscular mycorrhizal fungi and are not pathogenic, so will not be discussed further in this text.)

- Ascomycota (the sac fungi). This is by far the largest phylum, with around 65,000 known species, and it is also the most ubiquitous group, as these fungi are found in all habitats, including (uniquely) the marine environment. The yeasts also belong to this phylum, as do many species that constitute the fungal component of **lichens**. The Ascomycota are named for the sac-like structures called **asci** in which they produce spores during sexual reproduction. The great diversity of the group (which consists of many classes, with many orders, each containing many families of genera) is reflected in its capacity for pathogenicity. Examples include *Claviceps purpurea* (ergot) (described in Research Box 1.1 in Chapter 1), the *Erysiphe* species that cause some of the powdery mildews of many plant species, and the *Sclerotinia* species that cause watery soft rots in vegetables. The list is almost endless, but suffice it to say that diseases affecting all parts of the plant can be caused by fungi in this phylum.

  Note: There are some species belonging to the Ascomycota that never reproduce sexually. These are called the Deuteromycetes, also known as the imperfect fungi or **mitosporic fungi** (because they only reproduce by mitosis), and they are sometimes classified separately.

- Basidiomycota (the club fungi). This is another large group, with about 30,000 species identified to date. These fungi are distinctive in appearance due to the formation of their familiar fruiting bodies, such as mushrooms and brackets, in which they produce sexual spores. Although the group is largely made up of **ectomycorrhizal** fungi (those that form a **mutualistic** and **symbiotic** association with many plants, especially trees), many species in the phylum also have a unique capacity to break down **lignin**, a tough polymer component of the plant cell wall that is particularly abundant in trees. Therefore one of the main food sources for these fungi is rotting wood, and saprophytic bracket fungi are commonly seen growing in abundance on dead trees and fallen branches. The vast majority of the fungi in this phylum are thus either mutualistic or saprophytic. However, a few genera are pathogenic, causing rusts and smuts, particularly in cereal crops. Examples of these include *Ustilago* species, which cause smuts in wheat and barley, and *Puccinia* species, which cause rusts in cereals. In addition, there are a few genera that cause tree diseases (as a consequence of the ability to break down lignin). Examples of these include *Armillaria* species that cause root rots in many trees, including fruit trees, and *Lenzites* species that cause brown rot in coniferous trees.

Table 2.1 provides a simplified overview of some of the fungal families that include important plant pathogens, with examples from each family.

Table 2.1 **Examples of fungal pathogens of plants.**

Plant Fungal Family	Examples			Notes
	Genus	Example Species	Disease	
Clavicipitaceae	*Claviceps*	*purpurea*	Ergot in grasses	The most well-known host is rye (see Research Box 1.1 in Chapter 1)
Erysiphaceae	*Erysiphe*	*cruciferarum*	Powdery mildew of brassicas	Particularly devastating to the Christmas-dinner favorite, Brussels sprouts
Glomerellaceae	*Glomerella*	*graminicola*	Cereal stalk rot (anthracnose)	Economically important disease of many cereals, especially maize
Mucoraceae	*Rhizopus*	*artocarpi*	Jackfruit fruit drop disease	Although many species of this genus are saprophytic, a few cause disease
Nectriaceae	*Fusarium*	*oxysporum*	Panama disease of bananas	Currently one of the most devastating banana diseases; it has wiped out a number of banana cultivars
Olpidiaceae	*Olpidium*	*brassicae*	Brassica root burn disease	Recognized more as a virus vector than as a direct disease agent
Physodermataceae	*Physoderma*	*alfalfa*	Alfalfa crown wart disease	Soilborne fungus that can lie dormant for many years and is therefore difficult to eradicate
Pleosporaceae	*Alternaria*	*arborescens*	Tomato stem canker	A vast genus of ascomycetes that cause a wide range of plant diseases
Pucciniaceae	*Puccinia*	*triticina*	Wheat leaf rust	Causes major commercial yield losses, the spores overwinter, and there are ongoing infection issues
Venturiaceae	*Apiosporina*	*morbosa*	Black knot disease of drupes	Forms large black structures that are often used as habitat by social insects. The disease is relatively easy to manage

2.4 Bacteria

Introduction

It is not known how many bacterial species exist worldwide, but estimates are in the tens of millions. This is not helped by the fact that bacteria reproduce, develop, and evolve so rapidly that any educated guess would become out of date almost immediately. We do know that the vast majority of bacteria are non-pathogenic, existing symbiotically or saprophytically. Of those that are pathogenic, surprisingly few species are associated with plant infections. It is estimated that around 120–250 bacterial species are plant pathogens. This is a significantly wide numerical range of estimates given the small numbers involved. However, it is generally only those pathogens of economic importance, such as *Pseudomonas* species, *Xanthomonas* species (for example, *X. campestris*; see Figure 2.5), and *Erwinia* species, that have been studied in any depth.

Bacterial Structure

The domain Bacteria is the main group of organisms known as prokaryotes. (The domain Archaea is also prokaryotic, as are the Tenericutes—the mycoplasmas and spiroplasmas of the domain Bacteria discussed later in this chapter in the section on other plant pathogens). Prokaryotes are defined as lacking true membrane-bound cell organelles, and most notably their DNA is not organized into a nucleus. They have a complex cell wall structure (discussed below), which may or may not be enclosed in a **capsule**—a sticky outer layer of either proteins or polysaccharides that aids adherence to host surfaces and/or offers resistance to **phagocytosis** by the host cell. The plasma membrane is the layer that separates the cell wall from the cell contents, and in many cases it may develop into folded regions (known as invaginations) that can perform specific cellular functions (for example, as sites for nutrient and waste transport), facilitated by the **periplasmic space** which is located between the cell wall and the plasma membrane, and contains hydrolytic enzymes. These regions of the plasma membrane may break away from the main layer to form structures called **mesosomes**, which are membrane bound although they are not true organelles.

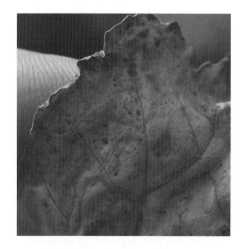

Figure 2.5 *Xanthomonas campestris* **pv.** *campestris* **(cabbage black rot) on a cabbage leaf.**

Figure 2.6 Structure of a generalized bacterial cell.

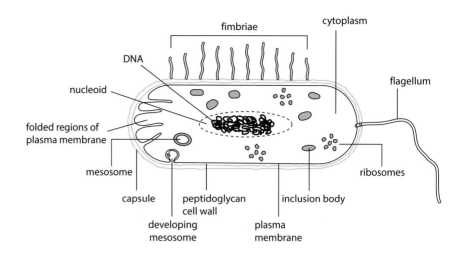

Within the cell is cytoplasm that contains DNA. The latter is commonly found in a less dense region within the cytoplasm, known as the **nucleoid**. Also within the cytoplasm are **ribosomes** for protein synthesis, as well as **inclusion bodies** for storage of metabolic products. **Plasmids** may or may not be present. These are semi-independent structures that contain their own chromosomal DNA. Although their functions have frequently been identified (for example, *Agrobacterium tumefaciens* contains **plasmids** that induce tumors in the plant host), in many species they have not yet been determined. They generally replicate synchronistically with the bacterial cell. Bacteria replication usually occurs by binary fission. However, many bacteria have small structures on their surfaces called sex **pili** that cause two cells to adhere to each other prior to DNA transfer, in a process known as **conjugation**. In addition, other pili called **fimbriae** may be located on the cell surface. These have the function of aiding attachment to surfaces. Other significant structures that may or may not be present include flagellae, which allow directional movement of the cells (Figure 2.6).

Bacteria come in a variety of shapes (spherical, rod-like, spiral, or irregular, also known as **pleomorphic**), sizes (0.5–10 μm), and nutritional types (ranging from **photoautotrophic** to **chemoautotrophic**).

The Bacterial Cell Wall

The intricate cell wall structure of bacteria includes **peptidoglycan**, a complex of sugars and polypeptides. The peptidoglycan layer is either a thick outer layer surrounding the plasma membrane or a thinner layer sandwiched between an outer lipopolysaccharide membrane and the inner plasma membrane (Figure 2.7).

Classification of Bacteria

Gram staining: There are a number of methods for characterizing bacterial species, and alongside the universally used binomial system, differentiation between the two cell wall types provides a further level of categorization. If bacteria are stained with two different specific dyes, namely crystal violet dye (which as its name suggests is violet in color), and safranin (a pinkish-red dye), the safranin dye stains all the bacteria, but the violet dye only stains bacteria that have a thick peptidoglycan wall. As the violet dye is more intense than the safranin dye, it masks the red color of the latter. In this case the bacteria are described as Gram positive (for example, the Actinobacteria), and the violet color is distinctive when viewed under a microscope. However, if the bacterial cell wall has an outer membrane, the violet dye can be flushed away leaving only the red dye. In this case the bacteria are described as Gram negative (for example, the Proteobacteria). Distinguishing between the two

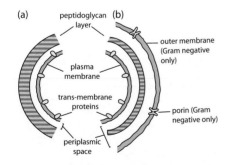

Figure 2.7 Bacterial cell wall structure. (a) Gram positive is simpler, with a wide peptidoglycan layer and the plasma membrane separated by the periplasmic space. (b) Gram negative is more complex, with a thin peptidoglycan layer within the periplasmic space, and is sandwiched between two membranes, with structural pores formed from transcellular proteins (known as porins). The capsule (not shown) may or may not be present.

types of cell wall structure can be an important first stage in developing strategies to combat bacterial infection, as the more complex cell walls of the Gram-negative bacteria can often render these pathogens more resistant to host defenses and to externally applied treatments. However, it should be noted that, as with most biological systems, there are exceptions to the rule, and some bacteria appear to be Gram neutral.

A further level of characterization that is used specifically for pathogens is the **pathovar** (pathogenic variety, often abbreviated to pv.). Pathovars differ solely in the host that they infect. For example, when *Xanthomonas campestris* causes black rot in cabbage leaves (Figure 2.5) it is known as *X. campestris* pv. *campestris*. However, when the same species infects rice, causing leaf streak disease, it is known as *X. campestris* pv. *oryzicola*, and when it infects cereals causing blight stripe it is known as *X. campestris* pv. *translucens*. Furthermore, a pathovar is specific in that it does not infect another pathovar's host.

Systems for Bacterial Taxonomy

Gram staining is one of the most commonly used methods of identifying bacteria, but it only separates them into two groups based on the composition of their cell walls, and there are a few bacteria that do not appear to conform to either state. As bacteria are submicroscopic, simple observations under the microscope only reveal such characteristics as shape (for example, the spherical shape of the cocci, or the rod-shaped bacilli). In addition, cell surface appendages, namely flagella or fimbriae, may or may not be present.

More recently it has become possible to identify bacteria using 16S ribosomal RNA sequencing. This gives a measure of relatedness between species. Another technique that is commonly used for all organisms is to measure the ratios of guanine and cytosine (G + C) in the DNA, as these tend to be much more variable in bacteria than in other organisms. However, in order to achieve conclusive taxonomic identification, full DNA genomic sequencing is required, and modern molecular techniques have significantly improved our understanding of the organization of the domain Bacteria. A simplified phylogeny of bacteria that specifically highlights some of the plant pathogens is shown in Figure 2.8.

Table 2.2 provides a simplified overview of some of the bacterial families that include important plant pathogens, with examples from each family.

Figure 2.8 Simplified phylogeny of bacteria down to the level of genus. Degrees of relatedness can be seen here. For example, *Erwinia* species and *Pseudomonas* species are related up to class level before they diverge, whereas *Clavibacter* species and *Streptomyces* species are related up to order level before they diverge. (Note that the list is not exhaustive.)

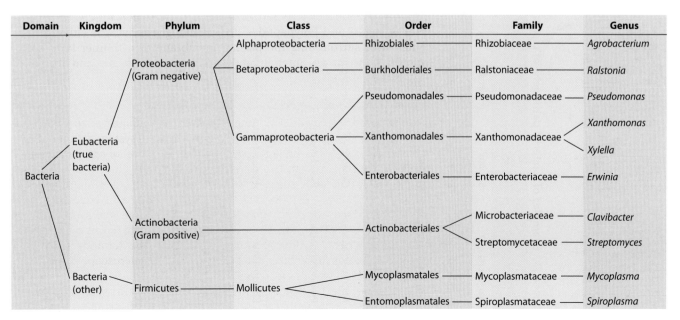

Domain	Kingdom	Phylum	Class	Order	Family	Genus
Bacteria	Eubacteria (true bacteria)	Proteobacteria (Gram negative)	Alphaproteobacteria	Rhizobiales	Rhizobiaceae	*Agrobacterium*
			Betaproteobacteria	Burkholderiales	Ralstoniaceae	*Ralstonia*
			Gammaproteobacteria	Pseudomonadales	Pseudomonadaceae	*Pseudomonas*
				Xanthomonadales	Xanthomonadaceae	*Xanthomonas*
						Xylella
				Enterobacteriales	Enterobacteriaceae	*Erwinia*
		Actinobacteria (Gram positive)		Actinobacteriales	Microbacteriaceae	*Clavibacter*
					Streptomycetaceae	*Streptomyces*
	Bacteria (other)	Firmicutes	Mollicutes	Mycoplasmatales	Mycoplasmataceae	*Mycoplasma*
				Entomoplasmatales	Spiroplasmataceae	*Spiroplasma*

Table 2.2 Examples of bacterial pathogens of plants.

| Plant Bacterium Family | Examples | | Disease | Gram Staining | Notes |
	Genus	Example Species			
Bacillaceae	*Bacillus*	*pumilus*	Mango blight	Positive	Many species are pathogenic, but also many species are used to control other plant diseases
Burkholderiaceae	*Burkholderia*	*caryophylli*	Wilt; carnation root rot	Negative	Species range is highly diverse, and ecological niches are equally diverse and ubiquitous
Enterobacteriaceae	*Erwinia*	*amylovora*	Fire blight	Negative	The common name of the disease refers to the blackened leaves. The disease mainly affects apples and pears, along with some other members of the Rosaceae family; other *Erwinia* species affect more diverse hosts
Enterobacteriaceae	*Pectobacterium*	*atrosepticum*	Potato black leg	Negative	There are only a few species, but diverse hosts. The pathogen causes a number of soft-rot-type diseases
Microbacteriaceae	*Clavibacter*	*michiganensis*	Wilt, canker, or ringspot (depending on subspecies)	Positive	Only one species is currently classified in this genus. Each subspecies is highly specific to host type
Pseudomonadaceae	*Pseudomonas*	*syringae*	Blister spot	Negative	Extremely diverse host range. The pathogen increases host vulnerability to cold weather by inducing ice nucleation at higher than normal temperatures, thus causing blisters where nutrients are more readily available
Rhizobiaceae	*Agrobacterium*	*tumefaciens*	Crown gall disease	Negative	Transmitted by Ti-plasmid, often used as a genetic tool
Streptomycetaceae	*Streptomyces*	*scabies*	Potato scab	Positive	Causes scabs and lesions mainly in potatoes and root crops. Other species cause various diseases in other plant hosts, such as cereals
Xanthomonadaceae	*Xanthomonas*	*campestris*	Wilts, spots	Negative	The overall host range is vast, although the pathogen is divided into pathovars that are host specific
Xanthomonadaceae	*Xylella*	*fastidiosa*	Chlorosis, scorching, lesions	Negative	The main hosts are citrus, drupes, olives, and grapes, although the pathogen can affect numerous other warm-climate crops, such as coffee and sunflowers, as well as common trees such as oak

2.5 Viruses

Introduction

The academic debate continues as to whether or not viruses can actually be characterized as living organisms—this is all about defining the criteria for life. On the one hand, it can be argued that viruses do not have a cellular structure, do not grow and develop, and are incapable of reproduction in the sense that they do not directly generate offspring. On the other hand, it can be argued that they are organic, they contain nucleic acid, and they do reproduce, albeit by using their host as a vehicle for their replication cycle. Whatever conclusions these debates may reach, the simple facts can be summarized as follows. Viruses are indeed organic, they do replicate when the host environment is favorable, they are often highly infectious, they cause significant diseases, and they are notoriously difficult to eradicate *in vivo*. It is clear that viruses do not fully conform to the accepted criteria for life, and therefore the standard method of taxonomic classification and the binomial system of nomenclature are generally not applied. The classification of viruses is thus commonly based on their physical characteristics. That said, a number of virologists do use a similar system of nomenclature to that utilized for other organisms, and it may be that this will become the standard in the future. Like all biological topics, this is a fluid and constantly developing field, and students should ensure that they keep up to date.

Nature's Own Genetic Engineers

Because viruses are so small, the nucleic acid complement only constitutes at most a few hundred nucleotides, and there is simply not enough nucleic acid for self-replication to occur. Instead, viruses re-encode host genetic material to produce new virus particles. As a result of this activity, they frequently disrupt the host cells' natural functions, causing physical deformities and biochemical changes. In addition, they are in a position to utilize the host's own cell products. However, the viruses themselves do not grow, they do not consume the host cells, and they do not produce toxins. These processes are discussed further in later chapters.

Viral Structure

Structurally, individual whole viruses are classed as particles known as **virions**. These range in size from around 20–400 nm in diameter, which is on average approximately 1% of the diameter of a bacterial cell (although they can extend to over 1 μm in length), and they are made up of nucleic acid surrounded by a protein **capsid**. Many virions also have a coat of lipids and/or carbohydrates, and sometimes other proteins. These are known as enveloped virions, whereas those that lack such a coat are termed naked virions. The capsid may be composed of one or several different types of protein; these are formed of repeating copies known as **protomers**.

Virions come in a variety of shapes (for example, circular, paired, oval, rod-shaped, filamentous, icosahedral, and a number of other less uniform shapes), and some have peripheral structures such as tails (Figure 2.9). During infection and replication within the host cell they exist as a disassembled mass of their components, and after replication they reassemble into whole virions prior to release from the host.

The nucleic acid in viruses is either DNA or RNA, but never both, and it can be double stranded (ds) or single stranded (ss). Single-stranded RNA viruses can be further classified as positive (+; identical to mRNA) or negative (–; complementary to mRNA). Single-stranded DNA is positively stranded and replicates via a double-stranded intermediate. Viral replication will be discussed further in Chapter 3. These characteristics form the basis of the Baltimore classification system for viruses.

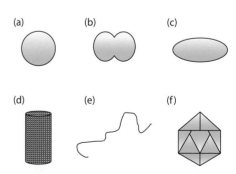

Figure 2.9 Examples of the variety of different shapes of viruses. (a) Circular (for example, chromovirus); (b) paired (for example, geminivirus); (c) baciliform (for example, rice tungro virus); (d) helical rod (for example, tobacco mosaic); (e) filamentous (for example, potato virus Y); (f) icosahedral (for example, cauliflower mosaic virus).

Classification of Viruses

The Baltimore Classification

Table 2.3 shows the Baltimore classification groups together with examples of plant pathogens belonging to these groups. The majority of plant viruses are in Group IV—that is, ssRNA(+).

However, other characteristics are also used to classify viruses, including the protein capsid. This is made up of either a single protein or sometimes a few different proteins. In both cases these are structured as continuous repeating copies (known as protomers). Thus capsid protein types, capsid symmetry, virus size, and the presence or absence of peripheral structures are also used as classification tools.

Viruses are further categorized, and subsequently named, by the morphological features that are expressed in the infected host. For example, mosaic viruses such as tobacco mosaic virus (TMV) (Figure 2.10) are categorized by and named for the mosaic-like discoloration and lesions that form on the host leaves. The three- or four-letter nomenclature, such as TMV in this case, is commonly utilized for viruses, and will be used throughout this text.

There have been moves by some virologists to place groups of similar viruses into orders, families, genera, and sometimes even species. In order to achieve this, the International Committee on Taxonomy of Viruses (ICTV) was set up in the 1970s. In 2011 they published their Ninth Report on Virus Taxonomy, in which they identified 94 families. This certainly represents the most comprehensive system in place at the time of writing, although there is still a long way to go. Table 2.4 provides a simplified overview of some of the virus "families" that include important plant pathogens, with examples from each family. It should be noted that numerous viruses have yet to be assigned to this taxonomic system, including some that are of significant economic importance, such as tobacco mosaic virus and rice stripe virus, which are listed separately as "unassigned" at the end of the table.

Students will find many different virus classifications systems in the literature, and should be aware that this is an ever-changing and much debated field. It is therefore important to keep up to date on this topic.

Figure 2.10 Tomato leaf infected with tobacco mosaic virus (TMV), showing mosaic mottling effect and necrosis of tissues.

Table 2.3 Baltimore classification of viruses, with examples of plant pathogens.

Group	Nucleic Acid Structure	Example of Plant Virus		Virion Shape
I	dsDNA	Chloroviruses		Icosahedron
II	ssDNA(+)	Geminiviruses (for example, African cassava virus)		Paired—incomplete
III	dsRNA	Wound tumor virus (WTV)		Icosahedron
IV	ssRNA(+)	Tobacco mosaic virus (TMV)		Rod (helical)
V	ssRNA(−)	Citrus psorosis virus (CPV)		Filamentous
VI	ssRNA(+) RT[a]	Chromoviruses		Circular
VII	dsDNA RT[a]	Caulimoviruses	Rice tungro bacilliform virus	Bacilliform
			Cauliflower mosaic virus	Icosahedron

[a]Retrovirus.

Table 2.4 **Examples of plant viral pathogens.**

Plant Virus "Family"	Examples		Nucleic Acid Structure	Transmission Vector or Method
	Genus	Disease		
Betaflexiviridae	*Vitivirus*	Grapevine viruses A, B, D, E, and F	ssRNA(+)	Aphids, mealybugs
Bromoviridae	*Cucumovirus*	Cucumber mosaic virus	ssRNA(+)	Insects, mainly aphids
	Ilarvirus	Tobacco streak virus		Inherited via seed; pollen
	Bromovirus	Brome mosaic virus		Beetles
Bunyaviridae	*Tospovirus*	Tomato spotted wilt virus	ssRNA(−)	Thrips
Caulimoviridae	*Caulimovirus*	Cauliflower mosaic virus	dsDNA (RT)	Insects, mainly aphids
Circoviridae	*Nanovirus*	Subterranean clover stunt virus	ssDNA	Inherited (usually latent)
Closteroviridae	*Closterovirus*	Beet yellows virus	ssRNA(+)	Insects, mainly aphids
	Crinivirus	Lettuce infectious yellows virus		Whitefly
Comovirinae (subfamily)	*Nepovirus*	Tobacco ringspot virus	ssRNA(+)	Aphids, thrips, nematodes; seed
	Fabavirus	Broad bean wilt virus		Aphids
Geminiviridae	*Begomovirus*	African cassava mosaic virus	ssDNA	Whitefly
	Curtovirus	Beet curly top virus		Hoppers; contact with other plants
	Mastrevirus	Maize streak virus		Hoppers; grasses
Luteoviridae	*Polerovirus*	Potato leaf roll virus	ssRNA(+)	Contact with other plants
	Luteovirus	Barley yellow dwarf virus		Insects, mainly aphids; grasses
Partitiviridae	*Alphacryptovirus*	White clover cryptic virus 1	dsRNA	Inherited via seed
	Betacryptovirus	White clover cryptic virus 2		Inherited via seed
	Varicosavirus	Lettuce big-vein virus		Fungi
Potyviridae	*Bymovirus*	Barley yellow mosaic virus	ssRNA(+)	Fungi, grasses
	Potyvirus	Plum pox virus		Aphids
		Potato virus Y		Aphids with "helper virus"
		Tulip breaking virus		Insects, mainly aphids
	Tritimovirus	Wheat streak mosaic virus		Mites
Reoviridae	*Oryzavirus*	Rice ragged stunt virus	dsRNA	Insects, mainly hoppers
	Phytoreovirus	Wound tumor virus		Insects, mainly hoppers
Rhabdoviridae	*Nucleorhabdovirus*	Beet leaf curl virus	ssRNA(−)	Insects, beet lace bug
		Potato yellow dwarf virus		Insects, mainly aphids
Sequiviridae	*Waikavirus*	Rice tungro spherical virus	ssRNA(+)	Aphids, hoppers
	Sequivirus	Carrot necrotic dieback virus		Aphids
Tombusviridae	*Tombusvirus*	Tomato bushy stunt virus	ssRNA(+)	Unknown
	Carmovirus	Carnation mottle virus		Unknown
Tymoviridae	*Tymovirus*	Turnip yellow mosaic virus	ssRNA(+)	Beetles
Virgaviridae	*Pecluvirus*	Peanut clump virus	ssRNA(+)	Seed; fungi; protista
Unassigned				
	Tobamovirus	Tobacco mosaic virus	ssRNA(+)	Contact with other plants
	Tenuivirus	Rice stripe virus	ssRNA(−)	Grasses; hoppers

2.6 Other Plant Pathogens

Introduction

Any attempt to discuss the taxonomy of organisms under the heading "other plant pathogens" is challenging. There are many phyla that once sat quite comfortably within one of the original five kingdoms, but which

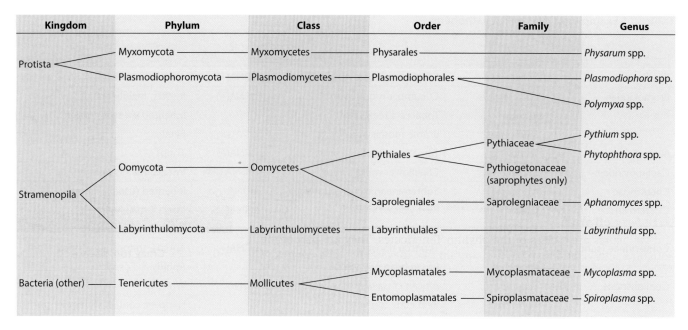

Figure 2.11 Simplified phylogeny of other plant pathogens—that is, those not found within the main taxonomic groups. (Note that the list is not exhaustive.)

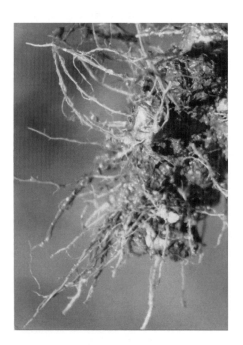

Figure 2.12 Club root caused by *Plasmodium brassicae* **on cauliflower root.**

have now been more accurately classified and placed completely differently. For example, there are a number of fungus-like phyla, such as the Oomycota, which includes the downy mildews and, together with the Labyrinthulomycota, was once considered to belong to the Fungi. These phyla then went through a transitional period of being placed in the kingdom Protista. They now belong to the kingdom Stramenopiles, along with a number of other phyla that appear quite different, such as the diatoms and the brown algae (Phaeophyta). The Oomycetes are featured widely throughout this text, as the group includes economically important pathogens such as the family Pythiaceae. Although in many ways they are very similar to the true fungi, the Oomycota are distinguished from the latter by the production of motile biflagellate zoospores, a feature that helps to facilitate their pathogenicity. There are also other fungus-like phyla that remain in the kingdom Protista, such as the Myxomycota, which include the slime molds (Myxomycetes), an ancient and interesting group with a reproductive cycle that includes both a unicellular phase and a multicellular phase. As well as the fungus-like pathogens, we shall also discuss here the protozoa and the **mollicutes**. To clarify these taxonomic groupings specifically in relation to plant pathogens, a simplified phylogenic tree is shown in Figure 2.11.

Fungus-Like Pathogens

Both the kingdom Protista and the kingdom Stramenopiles have plant pathogen representatives. The Protista contains an odd mix of organisms, many of which seem morphologically dissimilar (to the extent that many taxonomic systems have now abandoned the categorization of Protista as a kingdom). However, among these are a number of fungus-like plant pathogens from the phylum Myxomycota, most notably the slime molds, and the phylum Plasmodiophoromycota, most notably *Plasmodium brassicae*, which causes club root in brassicas (Figure 2.12), and *Polymyxa* species, which cause diseases in cereals.

Representatives from the kingdom Stramenopiles include members of the phylum Labyrinthulomycota, which are not a serious threat to plants although they have been linked to diseases in turf grass. However, the phylum Oomycota includes some of the most well-known and damaging plant pathogens, including *Pythium* species and *Phytophthora* species. These cause a whole range of damping-off, root rot, blight, and canker-causing diseases that will be discussed in much greater detail in later chapters. Certain members of this group also have some quite alarming effects on the host morphology, such as *Sclerophthora macrospora*, which causes crazy top disease in a number of plants, including maize (*Zea mays*) (Figure 2.13).

Pathogens in the Domain Bacteria (Non-Eubacteria) With No Cell Wall

The Mollicutes are a class of the phylum Tenericutes. They are part of the domain Bacteria, and many texts maintain that they belong with the rest of this domain, namely the Eubacteria. However, the Tenericutes are structurally quite different. The eubacteria have highly complex cell walls, and these structures are fundamental to their functionality, so here the class Mollicutes will be considered separately, as unlike the eubacteria they have no such cell wall. The three main groups of mollicutes are the acholeplasmas, the mycoplasmas, and the spiroplasmas. The acholeplasmas do not include any known plant pathogens at the time of writing, so will not be discussed further here.

Mycoplasmas have been recognized for many years as pathogens of humans and animals, but in the 1960s when similar organisms were recorded in plant phloem tissue, full identification was not carried out and they became known as mycoplasma-like organisms (MLO). The term "MLO" has stuck, despite our better understanding of their taxonomy, largely because through observations they are still very difficult to identify with certainty—they simply appear as a tiny membrane-bound mass. Mycoplasmas that affect plants are typically renamed phytoplasmas. For practical reasons in the field, and more specifically in the farming industry, they tend to be identified by their effects rather than by microscopy and DNA analysis. They are transmitted by insect vectors, and most of them reside fastidiously in the phloem tissue.

Phytoplasmas are responsible for a number of important crop diseases, notably in fruit trees such as apple, pear, and peach, although many also cause damage to root vegetables, especially carrots (Figure 2.14), and to horticulturally important plants. Many diseases caused by MLO are known collectively as "yellows", as they often cause yellowing of the plant. However, yellows also induce other symptoms, such as proliferative growth, necrosis, and stunting.

Spiroplasmas are a "breakaway" genus of mycoplasmas. They can be distinguished by their helical shapes, although they are often **heteromorphic**. The most significant disease caused by this group is citrus stubborn, which induces reduced leaf and fruit size and lopsided fruit development. These species also cause a number of other plant diseases.

Table 2.5 gives a simplified overview of some of the other plant pathogen families, with examples from each.

A Note on Nematodes and Nutrient Deficiency

Nematodes are parasitic organisms that are sometimes included in texts about plant pathology because the symptoms that they induce in plants,

Figure 2.13 Crazy top disease (*Sclerophthora macrospora*) **is the cause of unusual morphological changes in plants such as maize (*Zea mays*).** A color version of this figure can be found in the color plate section at the end of the book. (Courtesy of FX Schubiger, pflanzenkranheiten.ch)

Figure 2.14 Carrot (*Daucus carota* subsp. *sativus*) infected with a mycoplasma-like organism (MLO). Note the hair-like growths protruding outward throughout the length of the carrot. (Photo adapted courtesy of Harold F Schwartz, Colorado State University, Bugwood.org.)

Table 2.5 Examples of other plant pathogens.

Plant Pathogen Family	Example Genus	Example Species	Disease	Notes
Mycoplasmataceae	*Phytoplasma*	*asteris*	Aster yellows disease	Branch of the domain Bacteria, although significantly different in structure. There are many species and many hosts
Peronosporaceae	*Sclerotopthora*	*macrospora*	Crazy top disease of cereals and rice	An oomycete that affects numerous hosts, although mainly cereals. It induces outgrowths of the ears
Pythiaceae	*Pythium*	*aphanidermatum*	Turf blight	Once classified as fungi, now known as one of the oomycetes. There are numerous species, which affect most plants
Pythiaceae	*Phytopthora*	*hevea*	Pod rot and black stripe of rubber trees	As well as rubber, this pathogen affects many tropical trees, especially nut-bearing species
Trentepohliaceae	*Cephaleuros*	*virescens*	Coffee red rust	A small genus of algae, many of which cause disease in tropical plants

such as root damage, stunted growth, and lesions, can often be mistaken for pathogenic infection. For the sake of completeness, nematodes have been included in the compendium (see Chapter 9). The same situation applies to many symptoms caused by nutrient deficiency. For example, manganese deficiency can cause mottling of leaf color in many plants, which could easily be mistaken for viral infection (see Figure 1.5 in Chapter 1).

2.7 Summary

Desirable as it would be to think that the studies of molecular phylogeny that have precipitated the changes we have seen in taxonomy have reached some kind of zenith, and that all known species are sitting comfortably where they should be, the pace at which taxonomy has moved on in recent years leads us to advise students to keep well up to date with this topic. At the same time they need to remain aware that even today, indeed perhaps now more than ever, biologists are using numerous and varied systems, and therefore different texts may refer to different phylogenic placements of organisms.

In this book we have attempted to simplify this topic in order to help students to grasp the basic principles of taxonomy specifically with regard to plant pathogens, and the coverage is by no means exhaustive.

Further Reading

Agrios GN (2005) Plant Pathology, 5th ed. Elsevier Academic Press.

Coenye T & Vandamme P (2003) Diversity and significance of *Burkholderia* species occupying diverse ecological niches. *Environ Microbiol* 5:719–729.

Deacon JW (2006) Fungal Biology, 4th ed. Blackwell Publishing.

European Food Safety Authority (2013) EFSA issues urgent advice on plant bacteria *Xylella fastidiosa*. www.efsa.europa.eu/en/press/news/131126

Maki LR, Galyan EL, Chang-Chien MM & Caldwell DR (1974) Ice nucleation induced by *Pseudomonas syringae*. *Appl Microbiol* 28:456–459.

Plamann M (2009) Cytoplasmic streaming in *Neurospora*: disperse the plug to increase the flow? *PLoS Genet* 5:e1000526.

Prescott LM, Harley JP & Klein DA (1993) Microbiology, 2nd ed. Wm. C. Brown Publishers.

Talbot N (ed.) (2001) Molecular and Cellular Biology of Filamentous Fungi. Oxford University Press.

Urry LA, Cain ML, Wasserman SA et al. (2016) Campbell Biology, 11th ed. Pearson Education Inc.

Webster J & Weber RWS (2007) Introduction to Fungi, 3rd ed. Cambridge University Press.

Willey JM, Sherwood LM & Woolverton CJ (2017) Prescott's Microbiology, 10th ed. McGraw-Hill Education.

Chapter 3
Infection Processes

The evolution of plant life has taken place over 440 to 480 million years, and during this time plants have evolved myriad physiological and biochemical mechanisms that have enabled them to survive in the many and varied environments on Earth. These environments range from tropical wet habitats to desert, montane, and arctic environments, and each type presents a challenge to plant life. For example, in tropical wet regions the continuous rainfall (over 3000 mm annually) and the numerous fungal pathogens that exist in those regions might be expected to quickly kill off any plant life, but clearly this is not the case. In fact plant life in the tropics is lush and diverse—evidence that plants have evolved mechanisms for survival in these tropical wet regions. In drier environments (for example, both hot and cold deserts), the main challenge to plant life is retention of cellular water; in hot environments, water is rapidly transpired, and in cold environments it is often locked up as ice. In order to survive in such habitats, plants have developed an armory of morphological adaptations, such as a thick waxy cuticle, reduced leaf area, spines, hairs, and convolutions. All of these features help to maintain plant life by reducing herbivory (through the presence of the spines and hairs) and decreasing water loss by evapo-transpiration (via the thick waxy cuticle, which also helps to minimize pathogenic infection). Other mechanisms that aid plant life in deserts involve adaptations at the biochemical and physiological level, such as increased protein metabolism, accumulation of carbohydrates, and the evolution of Crassulacean acid metabolism (CAM), all of which act to maintain an aqueous cellular environment.

An early and key development in plant anatomy is the cuticle. The cuticle acts as a barrier that protects plant cells from desiccating climates and invasion by pathogenic organisms. In terms of protection from external deleterious biotic and abiotic agents, the cuticle is the first line of defense. The next barrier to these agents is the durable cell wall, which provides strength and maintains cell integrity. Plants have evolved several other forms of anatomical strengthening, such as vascular tissue, **secondary thickening**, and ultimately bark in the woody plants. A pathogen must negotiate all of these defenses in order to infect plant cells and complete its life cycle.

Plant pathogens have co-evolved with plant life. The fossil record reveals a vast array of plant–pathogen interactions from as early as the Lower Devonian, some 400 million years ago. Consequently, pathogens have developed numerous ways to overcome plant external defense mechanisms. These include, but are not limited to, the following:

- Direct entry through a wound or through natural plant openings such as stomata and lenticels

- Enzymatic degradation of cuticular waxes

- Transmission of infection via **phytophagous insects**, particularly in the case of the viral pathogens.

Upon entry into the cellular environment of a plant, pathogens have to successfully spread and colonize the plant tissues. Again plants are not defenseless, and numerous host resistance reactions (see Chapter 4) may isolate a pathogen. However, pathogens may quickly spread through the intracellular spaces and vascular system of a plant, resulting in rapid systemic infection and plant death. In order to fully appreciate the challenges to successful parasitism, it is necessary to review some important structural components of plants and then discuss how pathogens negotiate these mechanisms and successfully infect plant tissues.

This chapter will review the plant cuticle, plant cell walls, and the organization of plant tissues, including the vascular system. It will then explore how pathogens overcome these obstacles and how they successfully colonize plant tissues and ultimately establish successful cycles of parasitism.

3.1 Review of Plant Anatomy

In plants, the interface between the leaf surface and the atmosphere is mediated by the deposition of cuticular waxes, which form a hydrophobic barrier between the cellular environment of the plant and the atmosphere. The cuticle has several layers (Figure 3.1), each of which has a unique biochemical composition. The cell wall is the starting point. It is composed of the polymer cellulose, which consists of millions of glucose molecules cross-linked together to form cellulose microfibrils (Figure 3.2). The next layer is the actual cuticle, which is composed of cuticular waxes and carbohydrate polymers that extend from the underlying cell wall and help to maintain cellular integrity. The next layer is composed of cutin and intracuticular waxes, and finally there is a layer of epicuticular waxes covering the cutin layer.

The cuticle has two components. The first is the polymer cutin, a three-dimensional network of fatty acids laid down within the outer regions of the cell wall and in a layer covering the outer cell surface. The second component, cuticle wax, is laid down in the fatty acid network and also on the

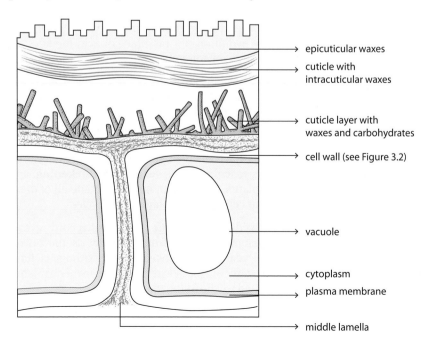

Figure 3.1 Anatomy of the plant cuticle showing layered structure.

epicuticular waxes

cuticle with intracuticular waxes

cuticle layer with waxes and carbohydrates

cell wall (see Figure 3.2)

vacuole

cytoplasm

plasma membrane

middle lamella

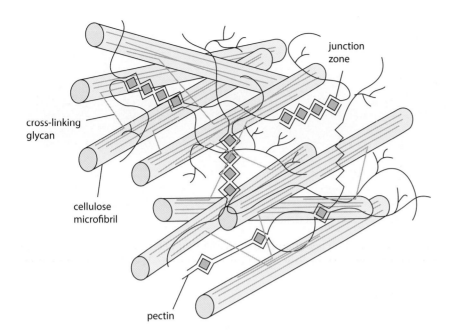

Figure 3.2 Anatomy of plant cellulose showing cross-linked microfibrils.

junction zone

cross-linking glycan

cellulose microfibril

pectin

outer cell surface above the matrix of fatty acids. The waxes are esters of fatty acids from the C_{16} and C_{18} families of acids, such as 16-hydroxy palmitic acid and 18-hydroxy stearic acid, and long-chain alcohols. The C_{16} and C_{18} acids are known as omega hydroxy acids (ω-hydroxy acids), which are a class of naturally occurring straight-chain aliphatic organic acids comprised of n carbon atoms, with a carboxyl group at position 1 and a hydroxyl group at the terminal n carbon atom. The complexity of cutin is partly explained by the presence of the carboxyl groups in the C_{16} and C_{18} fatty acids, as they can form bonds with the hydroxyl group of other fatty acids and the long-chain alcohols, which allows the formation of the branched structures that characterize the appearance of the epicuticle. The thickness of the cuticle varies depending on the species and on many environmental variables; for example, plants from a mesotrophic environment have a much thinner cuticle in their leaf tissue (0.1–1.4 μm) than plants from a xerotrophic environment.

The development of secondary thickening in plants is a major evolutionary step that has allowed plants to grow into huge structures, such as the **dipterocarp** trees of Asia, many of which exceed 75 m in height, and even taller trees such as the giant sequoia (*Sequoiadendron giganteum*), which can exceed 110 m in height. This is because secondary thickening provides internal and external supporting structures such as vascular tissue and bark, respectively. A range of different cell structures are involved in secondary thickening (Figure 3.3), and when organized into specific tissue types, secondary thickening develops into important vascular tissues such as the xylem and phloem (Figure 3.4). Xylem consists of a collection of pipes and tubes that run throughout the plant and facilitate the distribution of water and minerals from the roots to all the plant tissues. Phloem consists of a cluster of pipes and tubes running in parallel with the xylem, and is involved in the translocation of **photosynthates** from the leaves to all the other plant tissues. The evolutionary zenith of secondary thickening is the development of bark, which provides a very durable external surface that protects the thin living cambium layer from the numerous external agents that could impair the biological integrity of the plant. Bark contains numerous **lenticels**, which are essential for gas exchange between the cambium tissue and the atmosphere, but can also serve as a potential point of entry for pathogens.

Roots are biologically complex structures that both anchor the plant and facilitate the uptake and transport of water and mineral nutrients. The

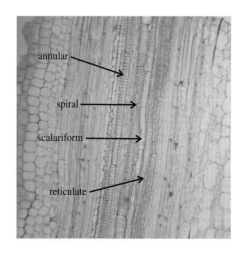

annular

spiral

scalariform

reticulate

Figure 3.3 Examples of secondary thickening tissue types in a eudicotyledenous stem. A color version of this figure can be found in the color plate section at the end of the book.

Figure 3.4 Vascular bundle tissue from a eudicotyledenous stem. A color version of this figure can be found in the color plate section at the end of the book.

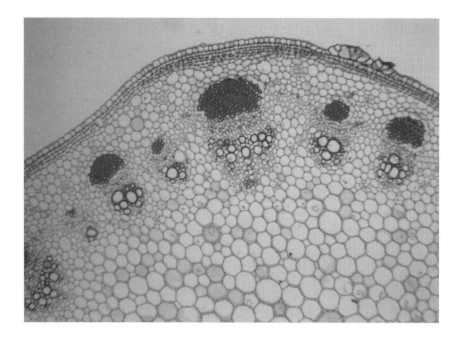

vascular tissue of the roots is continuous with that in the above-ground parts of the plant, and the root structure often also has secondary thickening as described above, which helps to support larger plants. In addition, roots are able to exert root pressure via osmosis, effectively pushing the water and minerals up through the xylem. Most plants enjoy a mutualistic relationship with mycorrhizal fungi that are intimately associated with their roots. The hyphae of ectomycorrhiza grow between and penetrate into the plant root cells. The fungi are able to utilize photosynthates manufactured by their plant partner, and in return the fungi are able to draw on mineral nutrients such as phosphorus from beyond the **root depletion zone** and outside the **rhizosphere**. Hyphae are much finer and faster growing than plant roots, so are able to quickly expand outwards from the rhizosphere to obtain minerals by passing easily through air spaces between the soil particles. In addition to such symbiotic relationships between plants and fungi, some plants, namely the legumes, have a symbiotic relationship with certain bacteria, mainly *Rhizobium* species. These bacteria fix nitrogen (N), which is required by the plant for protein and nucleic acid synthesis, from the atmosphere. Like the fungi, in return for this they are rewarded with photosynthates. Furthermore, roots play an important role in vegetative reproduction. In addition, some plants, particularly those that are epiphytic, have aerial roots with functions that include both physical support and water acquisition.

3.2 Structure and Anatomy of Fungal Pathogens

Pathogenic fungi are by far the most numerous infective agents in plants, so before we progress to the complex topic of the infection process it is important that students have a working knowledge of the general structure and anatomy of these pathogens. This is essential because many of the infection processes employed by fungal pathogens rely on the specific structure of the parasite in question. The vegetative growth of filamentous fungi normally occurs via thread-like cell structures called hyphae (Figure 3.5), which eventually aggregate to form a colony known as a mycelium (Figure 3.6), which can occur on the surface or within the host. The hyphae may be divided up into separate cells by wall-like structures known as septa (singular = septum), or they may be undivided, in which case they are described as aseptate (Figure 2.3). Reproduction can be either sexual or asexual, and in both cases this relies on the production of spores. In asexual spores the

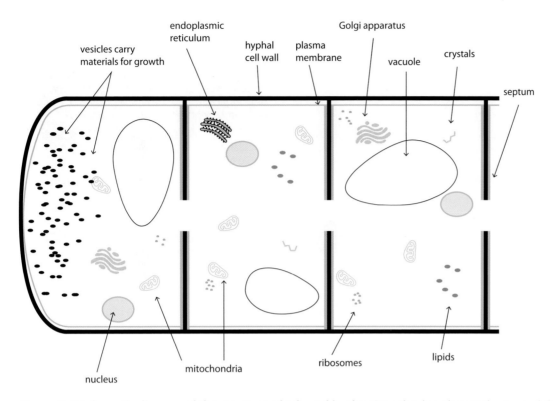

Figure 3.5 Schematic diagram of the structure of a fungal hypha. Growth takes place at the tip, and the materials required are transported to the tip in vesicles. Note that the cell compartments are only partially separated by the septa, thereby allowing easy movement of these materials. A color version of this figure can be found in the color plate section at the end of the book.

hyphae develop into a range of sporulating structures, of which the most common type is the conidiophore—a long chain of developing spores called conidia that originates from a hyphal branch (Figure 3.7). Conidia can also be produced directly from the hyphae. The conidiophores can be either simple subtending structures or they can aggregate into a compact coremium

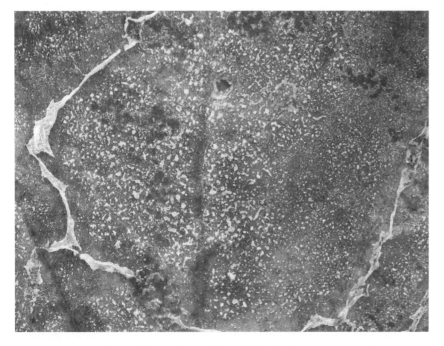

Figure 3.6 Fungal mycelia on the surface of a leaf of a courgette plant (*Curcurbita* species).

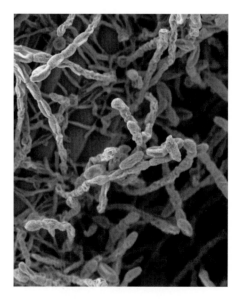

Figure 3.7 Conidiophore with conidia—the most common type of asexually produced hyphal spore.

Table 3.1 Summary of the range of fungal spores.

Spore Type	Dispersal Method	Sexual Stage	Key Fruiting Bodies	Resting Spore	Diagnostic Features	Example Crop Disease
Ascospore	Wind and rain splash	Sexual	Ascus and cleistothecium	No	In long chains, circular, oval, or elongated; sometimes septate	Powdery mildews from numerous genera (for example, *Blumeria, Erysiphe, Oidium*); numerous *Alternaria* diseases
Basidiospore	Wind and rain splash	Sexual	Basidium	No	Circular with a single nucleus	Rusts (*Puccinia graminis tritici*)
Chlamydiospore	Wind and rain splash	Asexual	Hyphal cell	Yes	One- to two- celled, thick-walled, round	*Fusarium* wilts, such as tomato wilt (*F. oxysporum* f. sp. *lycopersici*)
Conidium (pl. conidia)	Wind and rain splash	Asexual	Conidiophore from branching hyphae; a compact coremium, saucer-shaped acervulus, or flask-shaped pycnidium	No	Circular to oval; either light or dark in color; vary from one- to three-celled and can be filiform	Powdery mildews from numerous genera (for example, *Blumeria, Erysiphe, Oidium*)
Oospore	Wind and rain splash	Sexual	Antheridium attached to an oogonium	After fertilization	Thick-walled and spherical	Potato blight (*Phytophthora infestans*)
Sporangiospore	Wind and rain splash	Asexual	Sporangium	No	Spherical and dark colored	Soft rots of fruits and vegetables (*Rhizopus* species)
Teliospore	Wind and rain splash	Sexual	Telium	Yes	Thick-walled and two-celled; oval with hyphae attached on lower cell	Rusts (*Puccinia graminis tritici*)
Urediniospore	Wind and rain splash	Sexual	Uredium	No	Oval with two nuclei	Rusts (*Puccinia graminis tritici*)
Zoospore	Motile with flagella	Asexual	Zoosporangium	No	Variable with flagella	Clubroot (*Plasmodiophora brassicae*), damping-off (*Pythium* species), potato blight (*Phytophthora infestans*)
Zygospore	Wind and rain splash	Sexual	Development of progametangia from contact of zygospore tips before nuclei fuse; from germinating sporangium after resting	Yes	Circular to oval and augmented with needle-like projections; dark in color	Soft rots of fruits and vegetables (*Rhizopus* species)

Figure 3.8 Mummified apple resulting from infection by the brown rot fungus (*Monilinia* species).

or into very complex saucer-shaped fruiting structures such as an acervulus (plural = acervuli) or a flask-shaped pycnidium (plural = pycnidia), both of which erupt through the epidermis of the infected plant. There are many different types of spores (Table 3.1), and each type has diagnostic characteristics that are used to aid identification of plant disease. Sexual reproduction not only occurs via spores but can also result from the fusion of hyphae known as **dikaryotic** hyphae. This is where the nuclei of fertilized hyphae remain separate within the cells in pairs (N + N), and divides simultaneously to produce hyphae with pairs of nuclei. This process is observed in the ascomycetes and the basidiomycetes.

Hyphae can overwinter by forming a resistant resting stage known as a sclerotium (plural = sclerotia; from the Greek word *skleros*, meaning "hard"), which is a compact mass of hardened fungal mycelia. Such structures typify the resting stages of ergots (see Chapter 1, Research Box 1.1) of the Poaceae and the mummified apples of brown rot (*Monilinia* species) (Figure 3.8) and, following the onset of spring, the sclerotia will start a new cycle of growth.

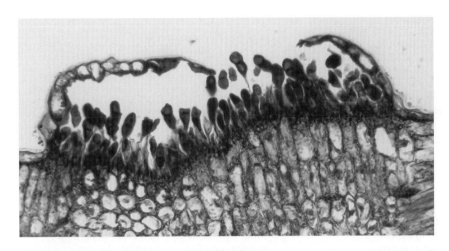

Figure 3.9 Teliospores, which are thick-walled two-celled sexually reproductive fungal spores that give rise to basidiospores.

The sexual spores produced by filamentous fungi, such as ascospores and basidiospores, increase the chances of survival of the fungus by producing new genetic variants. The spores and fruiting bodies of the sexual stage can survive on alternative hosts, such as the basidiospores of stem rust of wheat (*Puccinia graminis tritici*) that overwinter on the leaves and stems of barberry (*Berberis vulgaris* in Europe and *B. canadensis* in the USA). Basidiospores are produced from thick-walled two-celled teliospores (Figure 3.9), and thus give rise to complex life cycles (Figure 3.10). The complex nature of fungal life cycles is an essential element of the pathogenesis and epidemiology of plant fungal pathogens, and the student is advised to become familiar with the characteristics of fungal spores and their functions (Table 3.1).

3.3 The Rise and Evolution of Fungal Plant Pathogens

The close relationship between plants and their fungal pathogens is steeped in evolutionary history, and is associated with the development and evolution of biotrophic root-inhabiting fungal symbiosis. The fossil records illustrate one of the first known partnerships between plants and fungi, based on an endomycorrhizal fungal species of the genus *Glomites* that was isolated from Rhynite chert of the early Devonian period, around 400 million years ago. This relationship is based on the acquisition of mineral nutrients, and is closely linked to colonization of the terrestrial environment by plant life. The fossil record documents a relationship between the extinct plant *Aglaophyton major* (Figure 3.11) and the hyphae, chlamydospores, and arbuscules of a vesicular–**arbuscular** mycorrhizal (VAM) fungus belonging to the genus *Glomites*. The fossil record clearly demonstrates the infection process between these two organisms, where hyphae of *Glomites* are shown to be invading the intracellular spaces of *A. major* (Figure 3.12). Invasion of the plant begins along primitive root-like cells known as rhizoids. The mycelia of *Glomites* congregate along the axes of *A. major*, and penetration of cells occurs between the rhizoids via the swelling of the hyphae above the infection site. The swollen hyphal structure is postulated to be similar to an aspersorium used by extant biotrophic fungi such as the mildews. Following penetration of the plant cells, *Glomites* produces a range of hyphal structures that grow throughout the intercellular spaces of the plant and eventually form hyphal colonies in the cortical cells of *A. major*. These colonies subsequently develop into arbuscules in a row of thin-walled cells at the outer edge of the cortical cells of the host (Figure 3.12). Penetration of host cells takes place as relatively thick fungal hyphae squeeze through the intercellular spaces of the plant; this action produces thinner, elongated hyphal strands that subsequently penetrate the host cell wall. The host cells containing the arbuscules eventually form a dense aggregation that is highly

Figure 3.10 Life cycle of rust.

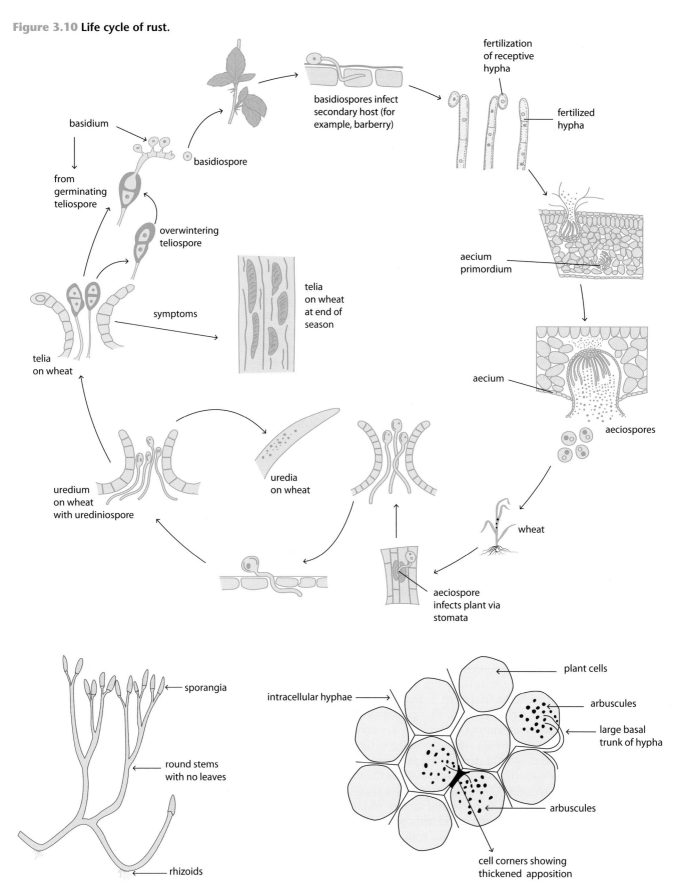

Figure 3.11 Schematic diagram of the extinct plant *Aglaophyton major*.

Figure 3.12 Schematic diagram of the hyphae of *Glomites* species plus the arbuscules within the cells of the *Aglaophyton major*.

dichotomous (bush-like). These arbuscule structures aid nutrient transfer between the host and the VAM fungus. Indeed, the fossil record documents the presence of *Glomites* species in all of the tissues of *A. major*, an evolutionary adaptation that makes sense as *A. major* had no leaves (Figure 3.11). The function of arbuscules is that of nutrient acquisition, which would of course be quite logical in plants growing in Rhynite chert from 400 million years ago, as this medium would have contained only limited nutrients such as nitrogen and phosphorus at that time, compared with later epochs. In such nutrient-limited soils, the next logical step for fungi was to develop the ability to be pathogenic, in order to exploit what little was available, and thus biotrophic fungi and fungus-like organisms such as the mildews and rusts may have evolved from these early mutualistic partnerships. A theory partly supported by the nutritional strategy of these pathogens is that they both form haustoria, which are one-way feeding structures that enhance the pathogen at the expense of the host.

This early record of biotrophic symbiosis illustrates a number of similarities to extant biotrophs, including documented plant responses to biotic challenge. In the *Glomites* example there is evidence that the plant responds to this biotic challenge by producing a thicker cell wall at the penetration site, which has been termed the collar. Conversely, internal colonies of *Glomites* avoid the cells of the hypodermis in *A. major*, because these cells contain **suberin**, but colonies freely establish in the cells of the cortex, where the suberin content is significantly reduced. In the case of extant biotrophic and necrotrophic pathogens, the host responds to pathogenic challenge by producing suberin (see Chapter 4). Finally, *Glomites* has morphological and functional similarities to the extant endophytic biotrophs such as the *Glomus* and *Sclerocystis* genera and other soilborne fungi that form obligate mutualistic associations with land plants. Many authors speculate that these early biotrophic relationships signify the co-evolution of plants and fungi. Indeed, these observations illustrate an important genetic model in plant pathology—the gene-for-gene concept (for a more detailed discussion of this concept, see Chapter 4).

The extensive timescales in the evolution of fungi have resulted in over 100,000 species, and during this time the early co-evolution between plants and fungi has given rise to a complex "arms race" between plants and their pathogens, to the extent that well over 8000 fungi that are known to be pathogenic have been documented.

3.4 The "Arms Race" Between Plants and Pathogens

The concept of an arms race is directly related to the gene-for-gene concept, but the philosophical question is how pathogens (which in this context include fungi, bacteria, and viruses) evolve and cope with change in their environment, and how this evolutionary adaptation gives rise to virulence with regard to parasitism.

Microorganisms adjust to environmental change at the genetic level, where changes in genes and the frequency of gene expression result in new biochemical, molecular, physiological, and morphological traits. These new traits confer an advantage on microorganisms in a changing environment, and a range of these traits have been postulated to confer a pathogenic advantage on microorganisms, in that a trait adapted for a new environment may also act as a virulence trait. This has been termed "dual-use traits." The following example helps to illustrate this hypothesis. Soil is a dynamic environment that is under constant biotic and abiotic challenge. For example, soils may freeze, dry out, and/or become waterlogged, or the rhizosphere may be modified by plant exudates such as **saponins** (that is, plant glycosides that have antimicrobial activity). Many microorganisms reside in soils and thus are constantly challenged by the above-mentioned factors, and one very damaging genus of soil bacteria is *Streptomyces*, which causes significant diseases in many root crops, such as potatoes. Soto et al. (2006) isolated a new virulence factor in *Streptomyces*, a gene homolog that encodes

Table 3.2 Summary of dual-use traits in plant pathogenic organisms.

Organism	Habitat	Disease and Crops	Dual-Use Trait	Role in Pathogenic Fitness	Putative Role in Environmental Fitness
Sclerotinia sclerotiorum	Soil and plant debris	Wet rot in broad beans, cauliflower, carrots, lettuce, hops, potatoes, and numerous other crops	Oxalate acid	Lowers cell pH	Secretion of oxalic acid in soil-dwelling ectomycorrhizal fungi has been shown to increase soil weathering and enhance nutrient uptake in associated vegetation, including phosphate and iron uptake
Rhizoctonia solani		Potato stem canker and black scurf; also numerous rotting diseases on many crop species, such as broad beans; survives saprophytically on plant debris		Helps polygalacturonase enzyme to penetrate cell wall; chelates calcium from middle lamellae; degrading enzymes and high levels of oxalic acid	As above, but saprophytic fungi have been shown to release oxalic acid which can bind calcium into calcium oxalate crystals and thus make calcium less liable to leaching
Sclerotium rolfsii		Damping-off and tuber, stem, and fruit rots in many crop species, such as brassicas, beans, and carrots		Highly virulent strains of *Sclerotium rolfsii* correlated with rapid growth rates of the pathogen and high levels of cell-wall-degrading enzymes and oxalic acid	Aids the degradation of lignin by basidiomycete fungi; several mechanisms are involved, but oxalate produced by wood-rotting basidiomycetes enhances lignin peroxidase and manganese-dependent peroxidase enzymes in white rot fungi Detoxifies substrates such as copper; oxalic acid binds soluble copper sulphates into insoluble non-toxic copper oxalate
Botrytis cinerea	Mainly on dead and dying plant remains	Causes gray mold on all plant parts of many important crop species	Efflux pumps	Confers advantage in pathogenesis by overcoming the plant defense phytoalexin resveratrol. Also confers resistance to the fungicide fenpiclonil	Confers resistance to soil antimicrobial compounds produced by soil microflora
Erwinia amylovora and *Agrobacterium tumefaciens*	Soil	Fire blight in pears and crown gall disease, respectively	Efflux pumps	Enhance resistance to plant antimicrobial compounds	Enhance resistance to soil flora and associated antimicrobial compounds
Gibberella species (also known as *Fusarium* species)	Soil and plant debris	Numerous rotting and wilting diseases in many crops	Mycotoxins such as trichothecene	Enhance virulence in pathogens, but have not been fully described	Mycotoxins act as antagonists to other microorganisms
Cladosporium fulvum	Soil, atmosphere, and as aerosols	Leaf mold in tomatoes	Virulence factors such as lysine motifs (LysMs)	LysMs are carbohydrate-binding protein molecules that are known to sequester chitin oligosaccharides which act as elicitors in host defense	May act as antagonist to microparasites

Data adapted from Dutton and Evans (1996) and from Morris et al. (2009).

a saponinase enzyme (that is, an enzyme that facilitates the breakdown of the host-derived saponin); this may be the result of adaptation to the rhizosphere, and hence enhance the virulence and pathogenesis of *Streptomyces*. There are numerous other examples of dual-use traits in plant pathogens (Table 3.2), and many of these traits have functions that are clearly aligned to virulence, such as **elicitors** in *Phytophthora infestans*. Elicitins in *P. infestans* induce the hypersensitive response (see Chapter 4) in plants, and a primary function of elicitin is the acquisition of sterols from the environment.

3.5 Infection Process of Plant Pathogens

Fungal Pathogens

Plant pathogens exhibit a wide array of strategies during the infection process. Direct entry may occur via wounds caused by herbivores, wind damage, and/or human activity, such as pruning. Fungi can also enter plants via their natural openings, such as the stomata and lenticels. However, in addition, fungi exhibit an array of complex biological interactions between host and pathogen. A number of infection strategies employed by pathogens require cell-wall-degrading enzymes to facilitate entry of the pathogen into plant tissues via degradation of the cuticle. Other strategies require the establishment of a pathogen–host interface via **haustoria** and the differentiation and function of other infection structures; many of these are related to the nutritional strategy (see Chapter 1) of the pathogen, with biotrophic and necrotrophic pathogens being two important and contrasting nutritional states.

Biotrophic Fungi

Biotrophic fungi such as the mildews and rusts depend on the integrity of the host cell for their survival, as these fungi manipulate the host's physiology for their nutritional benefit, so must establish successful colonization of the host without causing the death of host cells. However, in order to gain entry into the host, the parasite must negotiate the cuticle, and to achieve this it must first establish a presence on the external surface of the host. In the ubiquitous powdery mildews, the initial stage of infection is the attachment and germination of a pathogenic spore on a host, typically on a leaf or stem. This initial spore is often from windblown asexual conidia. Germinating conidia establish extensive hyphal growth that eventually forms mycelial mats on leaves, stems, flowers, and fruits (Figure 3.13). These mycelial mats produce sporulating hyphal growths, known as conidiophores in the mildews. The conidiophores produce vegetative spores known as conidia that contribute to successive cycles of infection in one growing season.

Spore attachment is mediated by a range of chemical compounds, which may be involved in downstream signaling during host penetration. Evidence of chemical mediation during spore attachment is provided by studies of a range of causative agents of mildew. In tomato powdery mildew (*Oidium neolycopersici*), studies have identified deposits of an extracellular matrix (ECM) beneath germination spores, the hyphae, **germ tubes**, and **appressoria** of *O. neolycopersici*, and around the edges of the penetration site. However, no such ECM has been observed under non-germinated spores. Similar ECM materials have been observed under germinated spores of *Blumeria graminis*, the organism that causes powdery mildew in the family Poaceae. It is speculated that the ECM aids spore adhesion and provides a medium for the localization of enzymes involved in penetration. Evidence to support such speculation is provided by binding assays that compared undifferentiated spores with the differentiated appressorial spores of *O. neolycopersici*. These assays concluded that the presence of an ECM was essential for adhesion, as undifferentiated and hence ungerminated spores could not remain attached to the host, whereas the appressorial stage could do so.

Figure 3.13 Mycelia of powdery mildew on strawberry (*Fragaria* species). A color version of this figure can be found in the color plate section at the end of the book.

In the powdery mildews, highly specialized infection structures develop from a conidial spore following attachment to the host's epidermis. The initial stage is the development of a primary germ tube that differentiates into an appressorial germ tube (Figure 3.14); this swells into an appressorium at the terminal end of the tube. The appressorium is the primary site of penetration for the pathogen, but strategies vary depending on the causative organism. In powdery mildews an **infection peg** grows from the appressorium, and through a combination of pressure and cell-wall-degrading enzymes (CWDE) such as cutinase, the cuticle is compromised and fungal hyphae can enter the host. The downward pressure excreted by appressoria is huge. For example, the appressorium of the rice blast fungus (*Magnaporthe oryzae*) can produce a turgor pressure of 8 MPa (80 bar), resulting in a drawing-pin effect and a ragged entry hole. Where CWDE act alone the entry hole is smooth-sided. However, most biotrophs use a combination of an infection peg and CWDE.

The differentiation of fungal tissue from spores to appressoria is mediated by cyclic adenosine monophosphate (cAMP). Evidence to support this hypothesis comes from experiments with barley powdery mildew (*Blumeria graminis* f. sp. *hordei*), where the distinct development of primary germ tubes and the emergence of appressoria is thought to be controlled by cAMP. The classic view of cAMP signaling involves cell-surface proteins that respond to a specific extracellular stimulus, which activates adenylyl cyclase (AC). This enzyme converts **adenosine triphosphate** (**ATP**) to cAMP via GTP-binding proteins. Changes in the activity of adenylyl cyclase result in changes in the concentration of cAMP, which then mediates extracellular signals to protein kinase A (PKA) by binding to its regulatory subunit. In barley powdery mildew, PKA transcripts were identified in developing conidia. To support these observations, experiments using exogenous applications of PKA and cAMP inhibitors were found to negatively affect germinating conidia. Adenosine cyclase and PKA have been identified and functionally characterized for a number of pathogenic fungi, and the cAMP signaling mechanism appears to play a crucial role in morphological differentiation of fungal tissue associated with germination and infection. In the rice blast fungus, AC **null mutants** showed a **pleiotropic** phenotype, including reduced germination and no appressoria.

There are numerous cell-wall-degrading enzymes. and each enzyme has evolved to overcome specific structures. For example, cutinase breaks down epicuticular waxes, pectinases or pectolytic enzymes degrade the pectin of the middle lamella, cellulases act on the cellulose chains, and hemicellulases degrade hemicellulose.

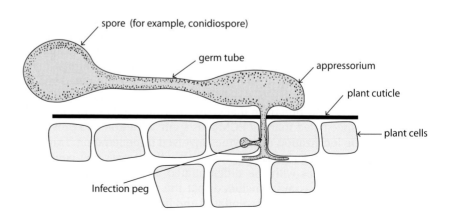

Figure 3.14 Schematic diagram of appressorium. (Adapted from Jones DG [1987] Plant Pathology: Principles and Practice. Open University Press.)

To gain entry to a host, the pathogen must initially breach the cuticle, and to achieve this, fungal pathogens produce cutinase, the enzyme that degrades cutin. Production of cutinase is controlled by a feedback loop, whereby the fungus produces a background level of cutinase, but following adhesion and initial degradation of the cutin into monomers, the fungus up-regulates cutinase production. This occurs within 15 minutes of the fungal spore being exposed to monomers of cutin. These monomers are detected by the fungal spore and thus activate the cutinase genes. Evidence of the genetic control of cutinase has been provided by studies using *Fusarium solani* f. sp. *pisi*, where expression of cutinase gene transcripts was observed to be maximal about 1 hour after inoculation of cuticle discs with the *F. solani* f. sp. *pisi* pathogen, and subsequently expression of the cutinase genes decreased. This correlated with detectable cutinase activity from 30 minutes to 45 minutes. In *M. oryzae*, cutinase production is controlled by a family of 14 to 17 genes. Three members of this family have been functionally characterized, and the gene coded as CUT2 is required for host surface sensing, normal formation of the infection structures, and complete virulence.

The next barrier that the pathogen must negotiate is the middle lamella, and this requires the production of pectinases to break down the pectin material of the middle lamella. Pectin is a complex polysaccharide that is composed of galacturonan molecules interspersed with rhamnose molecules and small side chains of galacturonan and other sugar molecules. Pectinases have specific actions. For example, pectin methyl esterase removes the small branches of the pectin chains, which alters the solubility of the pectin and enhances the rate at which pectin can be degraded by other classes of pectinases, such as polygalacturonases. Polygalacturonases add a water molecule to the pectin chain, thereby breaking the linkage between two galacturonan molecules. Pectinase lyses and then splits the chain by removing a molecule of water from the linkage, resulting in modified and smaller molecules. Each of the above-mentioned enzymes occur as endopectinases, which break up the pectin chain at random into shorter chains, or as exopectinases, which cleave the terminal linkage of the chain and thus release single units of galacturonan. The result of the action of these pectinases is the liquefaction of the pectin substances that hold plant cell walls together.

The control of pectinase activity during the infection process is mediated by the release of the galacturonan monomers or oligomers during the initial stages of infection (pathogens produce a background level of pectic enzymes). The galacturonan monomers and oligomers released by the initial infection are absorbed by the pathogen, resulting in enhanced production of pectin enzymes, and thus the galacturonan subunits act as effectors for pathogenesis.

In obligate biotrophic pathogens, secretion of CWDE does not result in total breakdown of plant cells. This would be counterproductive for this group of pathogens, as they need intact living plant cells for their nutrition. The fine control of these CWDE in biotrophs has not been fully described, but these destructive enzymes are immobilized by ionic interaction with plant cell wall exchangers. The electrical potential of the cell wall (ψ_{CW}) and the plasma membrane (ψ_{PM}) behave as ion exchangers, where the fixed cell wall charges interact with charged ions in the surrounding cellular matrix. Another mechanism that can reduce the activity of the CWDE is catabolite repression, whereby breakdown products of the cell wall (for example, galacturonan monomers) act as a feedback loop once they reach a critical threshold at the site of infection, and thus suppress any further synthesis of CWDE by the pathogen.

Following entry into the host cells, biotrophic pathogens develop a specialized feeding structure called a haustorium. Haustoria are specific to biotrophic fungi and fungus-like pathogens, and have a number of functions, including suppression of host defenses (Chapter 4) and the uptake and redirection of host-derived nutrients (Figure 3.15). Haustoria do not actually penetrate the host cells, but invaginate deeply into them, and a zone of separation is formed in between, forming an extrahaustorial matrix that is segregated by a modified host-derived extrahaustorial membrane. This membrane has been shown to have no ATPase activity, unlike the fungal membrane and the host membrane. The ATPase in the haustorial membrane acidifies the extracellular space, creating a pH gradient that drives sugar and amino acid–H^+ symport catalyzed by membrane-bound transporters in the haustorial membrane. This establishes a one-way traffic flow of nutrients (photosynthates) from the host to the pathogen. Photosynthates are absorbed by the haustoria and passed back through the hyphae to facilitate spore production. The spores themselves represent a carbon sink for the pathogen. Although it is not in the biotroph's best interest to cause excessive damage to the host, significant yield losses often result when carbon sinks become depleted. *Puccinia triticina* (formally known as *P. tritici*) typically infects wheat species (*Triticum* species). The pathogen utilizes carbon sink reserves designated for wheat grain filling, and although this pathogen does in fact stimulate enhanced nutrient production, modern commercial wheat species produce large grains and therefore have high nutrient requirements, and yield can be reduced by up to 50%. Clearly this is economically disastrous for the growers, who join the "arms race" on the side of the host plant.

Haustoria are not only required for nutrient uptake but are also involved in the process of infection by secreting **effector molecules**. Two lines of evidence support this theory. In the flax rust (*Melampsora lini*) it has been observed that cysteine-rich proteins are secreted from the haustorial complex and identified with avirulence. In this case the transport of the cysteine-rich proteins from the rust haustoria into the host cell suppresses host resistance mechanisms and hence the cysteine-rich proteins act as infection elicitors. The genes that control the expression of cysteine-rich proteins have been characterized, and correlate with the expression of the cysteine-rich proteins and lack of resistance reaction in the host. In flax rust these genes have been termed *Avr* genes (avirulence genes), namely *AvrL576*, *AvrM*, *AvrP4*, and *AvrP123*. Transient expression of *AvrM*, *AvrP4*, and *AvrP123* induced a hypersensitive reaction (see Chapter 4) in the host, and this was

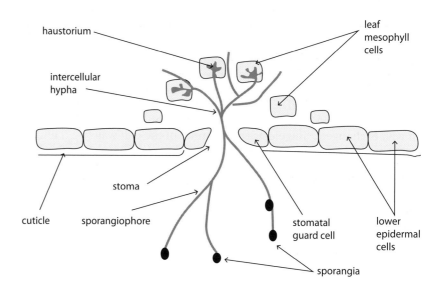

Figure 3.15 Schematic diagram of fungal infection by entry through leaf stoma, showing deep invagination of the mesophyll cells by haustoria.

haustorium

intercellular hypha

leaf mesophyll cells

cuticle   sporangiophore

stoma

stomatal guard cell

lower epidermal cells

sporangia

dependent on the corresponding resistance genes in flax (*M*, *P4*, and *P1/P2*), indicating that genes encode functional avirulence proteins. The complex *AvrP123* locus of the flax rust have corresponding flax resistance genes *P1*, *P2*, and *P3*, in *Avr123-A*, a cysteine-rich protein is encoded which is secreted into the host and blocks hosts protease activity. Hence *Avr123-A* mediates infection of host tissue. The second line of evidence to support haustorial-mediated pathogenesis comes from studies of the rust pathogen of broad bean (*Vicia faba*). Proteins from the haustoria of broad bean rust (*Uromyces fabae*) have been observed to be transported from the pathogen and then located in the host nucleolus. The secretion of these proteins is specific to the developmental stage of the rust fungus (that is, expressed from haustoria but not at spore stage). Again, a large array of genes (62 in total) are involved in the secretion of these extrahaustorial proteins in *U. fabae*.

Necrotrophic Pathogens

At first glance the infection strategies of necrotrophic pathogens appear less complex than those of the biotrophic pathogens, as they do not need to keep the host alive, since necrotrophs feed off the debris of dead plant tissue following infection. However, a closer inspection of the relationship between necrotrophic pathogens and their hosts reveals a staggering array of weapons used by necrotrophic fungi, such as CWDE, low-molecular-weight secondary metabolites, and toxic peptides.

Infection of plant tissue by widespread necrotrophic pathogens such as *Alternaria*, *Botrytis*, *Fusarium*, *Rhynchosporium*, and *Sclerotinia* proceeds via the formation of an inconspicuous appressorium and uniform infection hyphae, which grow and invade the intercellular spaces of the host. In contrast to the biotrophic pathogens, there is no requirement to keep the host alive. Thus necrotrophs start to kill the host via a range of strategies, including the secretion of CWDE, which cause significant tissue damage to the host. The destruction of host cells aids the infection process and provides the pathogen with carbohydrates. Indeed, there are many more genes encoding CWDE in necrotrophs than there are in biotrophs. For example, the genomes of *M. oryzae* and *Fusarium graminearum* contain 138 and 130 genes, respectively, that encode CWDE, compared with only 33 genes in the biotrophic corn smut fungus (*Ustilago maydis*).

Necrotrophs can also kill their host by secreting low-molecular-weight secondary metabolites such as the host-selective **HC-toxin**, a cyclic tetrapeptide produced by *Cochliobolus carbonum*, which is the causative agent of northern corn (maize) leaf spot. The HC-toxin is an essential factor in the infection process and the virulence of the pathogen. Race 1 of *C. carbonum* produces the HC-toxin, which results in large lesions on susceptible maize plants. Isolates of *C. carbonum* that are deficient in HC-toxin cause smaller and less damaging lesions on susceptible plants. HC-toxin acts as an inhibitor of histone deacetylases, and modifies plant gene expression patterns by histone hyperacetylation, which is thought to interfere with the transcriptional activation of host defense genes. Resistance to HC-toxin has been achieved by the breeding of maize plants that carry the *HM1-resistance* gene which encodes an HC-toxin reductase that detoxifies the HC-toxin. Another very virulent and toxic compound is the **T-toxin**, produced by *C. heterostrophus*, which is the causative agent of southern corn blight. This toxin is a linear long-chain **polyketide** with variants ranging from 35 to 47 carbon atoms. The toxic effects of the T-toxin occur in the host mitochondria, where the T-toxin binds with the host URF-13 protein, leading to pore formation in the inner mitochondrial membrane. This results in the leakage of molecules from the mitochondria, and the breakdown of the electrochemical gradient and disruption of ATP production in the host mitochondria. T-toxin synthesis in *C. heterostrophus* is controlled by three genes, namely *PKS1*, *PKS2*, and

DEC1. PSK1 and PSK2 encode two polyketide synthases, and DEC1 encodes an enzyme with decarboxylase activity. Deletion of any one of the three genes results in reduced virulence of the pathogen. The virulence of *C. heterostrophus* is manifested in maize plants that carry the Texas male sterility (TMS) gene, which leads to pollen sterility in maize. Virulence is strongly correlated with the T-toxin-producing isolates of *C. heterostrophus* (race T). Isolates of *C. heterostrophus* that do not produce T-toxin (race 0) display very weak virulence on all maize varieties, including TMS lines.

Another infection strategy that is used by necrotrophs is the production of toxic peptides, most of which are very specific to the host and are termed host-specific toxins (HSTs). However, a small number of toxic peptides are produced that are non-specific. One group of such non-specific toxic peptides produced by necrotrophic pathogens inactivate plant phytoalexins, which are toxic compounds produced by plants during host defense responses (see Chapter 4). Pathogens also produce general hydrolytic enzymes that inhibit plant hydrolytic enzymes but also act as toxins that induce chlorosis and necrosis in plants. *Rhynchosporium secalis*, the barley scald pathogen, produces a necrosis-inducing protein (NIP), which becomes active upon infection and modulates H^+-ATPase function in the host. All of the toxic peptides discussed above appear to be non-specific.

In contrast, other toxic peptides are host specific, such as toxins produced by *Pyrenophora tritici-repentis*, the causative agent of tan spot in wheat (*Triticum aestivum*). Races of *P. tritici-repentis* can produce multiple HSTs, of which at least two are proteins, namely the necrosis-inducing toxin Ptr ToxA and the chlorosis-inducing toxin Ptr ToxB. In addition to these two toxic proteins, a number of low-molecular-weight non-protein HSTs have been identified, including Ptr ToxC and Ptr ToxD. The virulence of *P. tritici-repentis* is strongly correlated with the copy number of genes that encode the above-mentioned HSTs. In the case of Ptr ToxA there is only one gene copy and the virulence of the pathogen is weak. However, the production of Ptr ToxB relies on multiple identical copies of the gene, and the virulence of *P. tritici* is strongly expressed. One further example of HSTs is the production of a complex system of proteinaceous HST compounds by *Stagonospora nodorum*, which causes leaf blotch in wheat. This fungal pathogen produces a range of HSTs termed SnTox1, SnTox2, SnTox3, and SnTox4. Four proteins encoded by distinct dominant host sensitivity genes (*Snn1*, *Snn2*, *Snn3*, and *Snn4*) are recognized by the SnTox proteins. The virulence of pathogens is thus strongly correlated with the gene-for-gene theory, and therefore fungal toxins matching specific host sensitivity proteins can be described in a one-to-one manner and consequently accounted for by the evolution of fungal races. Hence both a pathogen HST and a host susceptibility gene are required for a disease to occur via a direct interaction between the genes or indirect interaction of the gene products. This concept will be explored in more detail in Chapter 4.

3.6 Bacterial Pathogens

Introduction

Bacteria cause a number of plant diseases and can be responsible for significant economic losses through destruction of plant tissue or components of yield. They can produce a wide range of different symptoms, including leaf and fruit spots, damage to flower tissue (which is one of the early symptoms of fire blight of apples and pears, caused by *Erwinia amylovora*), mosaic patterns, necrosis, rots, and wilt disease (such as bacterial wilt of maize, caused by *Erwinia stewartii*). Bacteria can also cause hormone-based distortions of leaves and shoots, known as fasciations (caused by *Rhodococcus fascians*), and infection of tubers and stems can result in a range of symptoms, from tuber rots to whole-plant death (for example, crown gall disease, which is caused by *Agrobacterium tumefaciens*).

Bacterial pathogens commonly inhabit the soil, and soil residents are classified into two categories, namely soil inhabitants and soil invaders. A number of bacterial pathogens are true soil inhabitants, such as *Streptomyces scabies*, which causes common scab of potatoes. These soil inhabitants build up populations in the host plants and are able to persist in the soil on crop debris for 2 to 3 years. However, if susceptible hosts are grown in the same soil in successive years, the inoculum levels build up and only decline slowly following removal of the hosts. Many bacteria are less able to survive for long periods in the soil, and thus have been termed soil invaders, such as *Xanthomonas campestris* (the cause of blackleg in crucifers) and *Erwinia amylovora*. *E. amylovora* builds up its populations in the host and has developed persistent plant-to-plant infection cycles. In the case of apples and pears these hosts are long-lived perennial plants, and thus *E. amylovora* has lost the ability to persist in the soil, so its populations rapidly decline following removal of the host.

Soil bacteria of the genera *Azorhizobium*, *Bradyrhizobium*, *Frankia*, and *Rhizobium* develop symbiotic relationships with plants. These relationships involve the development of root nodules, which aid plant nutrition; a full description of root nodulation is provided in the compendium of diseases (see Chapter 7). Bacteria in the root nodules are able to fix atmospheric nitrogen. The development of root nodules is very similar to the development of arbuscules and haustoria in the fungi.

Infection Strategies

Like fungi, bacteria are able to enter the host via wounds and natural openings such as stomata and lenticels. Indeed, this is the main route of infection for plant pathogenic bacteria. Once inside the host, bacteria start to degrade plant cells by a range of CWDE, such as the pectinases (pectate lyase, pectin lyase, and pectin methyl esterase), cellulases, proteases, and xylanases. They then multiply rapidly, resulting in large colony-forming units (CFU) that continue the process of cell degradation via CWDE. In essence most plant pathogenic bacteria are necrotrophic. The exceptions to this are bacteria that form an association with plants, such as *A. tumefaciens* and root nodule-forming bacteria. Bacteria from these groups require adhesion mechanisms in order to recognize their host and facilitate aggregation in host tissues. These adhesion mechanisms range from production and export of glucan and the formation of pili in *A. tumefaciens* to other elicitor proteins that aid successful bacterial pathogenesis. The full life cycle and pathogenic process of *A. tumefaciens* are described in the compendium of diseases (see Chapter 7). Bacteria of the genera *Xanthomonas*, *Pseudomonas*, and *Xylella* also require an adhesion mechanism. This is because many of these bacterial organisms form in the dynamic environment of the xylem tissue and hence require physical augmentation of their cell surface for adhesion, such as the above-mentioned pili, and also effector proteins that interfere with the host immune response and consequently lead to pathogenicity. A good example is the type III secretion (TTS) system and associated pili observed in *Xanthomonas campestris* pv. *vesicatoria*.

It is essential for plant pathogenic bacteria (PPB) to be able to evade host defense mechanisms, and in order to achieve this, PPB have evolved a series of conserved secretion systems (observed in both mammalian and plant pathogenic bacteria) for successful colonization of host tissue. There are five types of secretion systems—type I-SS, type II-SS, type III-SS, type IV-SS, and type V-SS—each of which has specific roles to play in mediating pathogenesis in PPB (Table 3.3).

Table 3.3 Summary of secretion systems in plant pathogenic bacteria.

Secretion System	Function	Examples and Research Models (Not All are Plant Pathogens, Many are Pathogens of Fauna)
Type I	The simplest protein transport system. Involved in adhesion and pore formation, such as the secretion of α-hemolysin (HlyA) in *Escherichia coli* or metalloprotease of *Erwinia chrysanthemi*. The outcome is the development of membrane pores in the host cell	*Escherichia coli, Erwinia chrysanthemi, Pasteurella haemolytica*, and *Bordetella pertussis*
Type II	Targeting and assembly of membrane secretion. Secretion of enzymes into host cells	*Klebsiella oxytoca* and *Erwinia* species
Type III	A complex secretion system providing numerous layers of functionality as follows: (1) mediation of host contact; (2) energy mediation for protein secretion and translocation into host cell; (3) secretion and regulation of expressed genes encoding proteins secreted downstream; and (4) dedicated cytoplasmic chaperones for some secreted proteins	*Escherichia coli, Ralstonia solanacearum, Pseudomonas syringae*, and *Yersinia* species
Type IV	Transport of effector molecules, uptake and release of naked DNA, and propagation of genomic islands. Important in bacterial diversification and lateral mobilization of antimicrobial resistance and virulence genes	*Agrobacterium tumefaciens, Bordetella pertussis*, and *Helicobacter pylori*
Type V	A simple protein secretion mechanism; however, there are three components to the type V systems that aid infection, namely the auto-transporter protein and two partner secretion systems	*Escherichia coli*

3.7 Viral Pathogens

Infection by Viruses: Nature's Own Genetic Engineers

Viruses cannot survive away from their host or an intermediary carrier. They simply have no physiological means by which they can exist independently, and thus can only thrive by integrating themselves into host cell mechanisms. This extreme dependence means that infection of new hosts requires intimate contact. In a crop, where identical individuals are in close proximity to one another, this is an easy process that may involve touching of infected leaves, or intertwining of root systems between neighboring plants. Where potential hosts are more distant from each other, the most common method of transmission is through **vectors**—indeed the vast majority of plant virus transmission occurs via such means. Both methods of transmission are known as horizontal transmission. The other infection method, in which the virus is inherited from a parent host plant by its progeny either via the seeds or via asexual reproductive structures such as stolons, rhizomes, or bulbs, is known as vertical transmission.

Vectors

There are numerous and diverse insect vectors—for example, from the orders Diptera (true flies), Thysanoptera (thrips), Coleoptera (beetles), and Hemiptera (true bugs, such as aphids) (Research Box 3.1). Insects are by far the most common type of plant virus vector, but others do exist, such as nematodes (the roundworms) belonging to the phylum Nemathelminthes, plasmodiophorids (a parasitic group of Protista), obligate fungal parasite species of the genus *Olpidium*, and sometimes even other plants, such as the parasitic genus *Cuscuta* (dodder). As most vectors are able to travel between plant hosts, they are responsible for the large-scale spread of crop diseases. They will therefore be discussed in more detail in Chapter 5.

Virus Replication

Of course a single virion would have hardly any impact on its plant host, so the topic of viral replication is inseparable from the topic of viral infection. The transmission and spread of viruses within a host is entirely dependent on the host's own replication machinery. After initial passage into the host tissue, a full-blown infection can only result from virus replication.

Introduction

Aphids belong to the order Hemiptera (from the Greek *hemi*, meaning "half", and *ptera*, meaning "wing"). They do not actually have half wings, as implied by the name, although the forewings of many species in this order are divided into two physical textures—a hardened base and a membranous tip. The order Hemiptera is divided into three suborders, the Heteroptera, which are more commonly known as the true bugs, the Auchenorrhyncha, which consist mainly of the hoppers, and the Sternorrhyncha, which include the aphids. (The Auchenorrhyncha and the Sternorrhyncha are collectively termed the Homoptera, and these represent the majority of commercially important virus vectors.). There are around 4700 species of aphid, although less than 10% are commercially significant in terms of crop damage.

What makes the aphids, and a number of other species in this order, particularly relevant to plant pathologists is their highly evolved piercing and sucking mouthparts that allow them to tap directly into plant sap. This is damaging enough in itself, but it also makes them highly effective at transmitting plant viruses directly to and from their hosts in the process. In fact the Homoptera are responsible for an estimated 60–80% of all vector-transmitted viruses in plants. Moreover, in some cases the viruses can actually persist in the body of the vector for up to several weeks. This appears to have no adverse effect on the carrier, but it does give the virus an added advantage, as it can survive for longer while awaiting access to a new host. The possible disadvantage to the virus is that it is completely dependent on the feeding behavior of its vector. In addition, viruses have been detected in non-vector insect species, in which case they are unlikely to find a new target host plant.

Stylets: The Aphids' Feeding Straws

Aphid mouthparts consist of structures called **stylets** enclosed within a labial sheath, forming a rostrum or proboscis. The stylets (also sometimes known as **stomastyles**) (Figure 1) consist of four tubes arranged in two pairs—the outer mandibular stylets and the inner maxillary stylets. These structures penetrate deep into plant leaves and stems to extract nutrients from the phloem sieve tubes in the vascular tissue. The mechanism by which sap location and extraction occurs is still not fully understood. However, it is known that as the stylets penetrate through the layers of plant tissue the aphid pierces most of the cells that it encounters and samples the cell contents until it reaches the target tissue—the phloem sieve elements. Aphids produce a thick saliva that forms a gel, providing an intercellular sheath that gives each stylet physical protection as it penetrates deeper into the plant. They also produce a watery saliva that facilitates stylet penetration and that, during feeding, becomes mixed with the sap and gets drawn back up the stylets. This is thought to be a method of overcoming the plant's defenses by preventing coagulation of plant cell proteins designed to plug up cell damage.

Virus Survival Within the Vector

There are two fundamental types of viral transmission by vectors such as aphids, namely the cuticle-borne type (some-

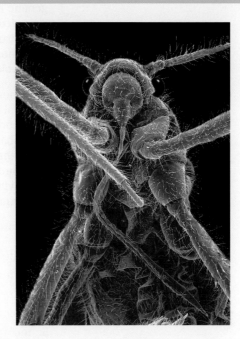

Figure 1 Close-up image of aphid showing the stylets. These are used for extraction of sap, but are also inadvertently the instrument of virus transmission. (Courtesy of Peter Smithers, Plymouth University.)

times known as the stylet-borne or non-persistent type), and the circulatory or persistent type.

The non-persistently transmitted viruses largely consist of the potyviruses (PYV) and a few other notable species, such as the cauliflower mosaic viruses (CaMV). Transmission between the vector and the plant host is rapid, as the virions cannot survive away from their hosts for more than a few hours, although usually the time to transmission is much shorter than this. They are retained for brief periods, mainly in the stylets but sometimes in other parts of the mouth, before the vector visits the next potential host. This short-term retention is facilitated by a virus-encoded protein that allows them to accumulate in protected sites.

The persistently transmitted viruses, such as cucumber mosaic virus (CMV) and potato leaf-roll virus (PLRV), need to circulate around the vector's body prior to transmission. Transmission is usually a slow process, with the vector having to reach deep into the host phloem tissue where the virions are obtained and later deposited. Virus particles are transported around the vector, specifically through the foregut, the hindgut, and the salivary gland via interactions with a series of proteins. They can survive within the vector for up to several weeks. Workers studying this process have been able to use these particular proteins as biomarkers to identify vector populations, which may ultimately enable agriculturalists to use biological control methods to manage pest numbers and thereby reduce viral infections of crop plants.

The salivary gland is the final port of call of the virus particles within the vector's body, and from here they are transported in the saliva back down the stylets prior to infection of a new host.

Viruses are so small that they do not have the capacity to contain sufficient nucleic acid to fully replicate independently. They are in fact nature's own genetic engineers, manipulating the host's genetic machinery to replicate new **viroids**. Release of these new viroids may or may not destroy the host cell. Viruses do not produce toxins, they do not grow, and they do not consume host cells. However, they do utilize host cell products via their replication cycle, and they disrupt cell processes.

(Another group of viruses exist, called satellite viruses. These are very unusual viruses that do not even have the normally fundamental capacity to re-encode host nucleic acid, possibly due to a defective gene. Instead they piggy-back other viruses known as helper viruses. However, they do replicate their own coat proteins. These viruses should not be confused with satellite RNA or satellite DNA, which also co-infect with helper viruses and cause virus-infective symptoms, but have no protein coat and are therefore constructed solely of nucleic acid.)

There are eight viral infection and replication stages:

1. Adhesion to the host cell
2. Formation of complexes with specific cell-surface protein receptors
3. Entry into the cell by endocytosis
4. Uncoating
5. Incorporation into the host genome
6. Viral genome replication
7. Maturation (assembly)
8. Release.

Each of these stages will now be described.

Adhesion to the host cell: The plant host cells have surface receptors for cell-to-cell communication. This mechanism is normally associated with biochemical signal transduction of a number of important cellular processes (for example, enzymes such as protein kinases required for stimulation of plant growth and development). Infecting virions can utilize these for attachment to the host cell wall and **formation of complexes with specific cell-surface protein receptors**. As we shall see later, the capacity to commandeer the host's normal systems and functionality is fundamental to virus survival. Once the virus has attached to the host cell surface it needs to be quickly taken inside the cell before the host's immunity mechanisms can be activated. **Entry into the cell by endocytosis** is a passive process. The cell membrane invaginates until it forms a separate breakaway unit (a vacuole) containing the virion. The virus protein coat layer disassembles in a process known as **uncoating**. This is followed by **incorporation into the host genome** in the cytoplasm by replacing similar host codons in a like-for-like manner. As many of the host genes encode similar proteins to those required for viral replication, only small numbers of host codons need to be exchanged to enable **viral genome replication**. Most plant viruses have an RNA genome and are largely comparative with (positive sense) or complementary to (negative sense) the host messenger RNA (mRNA). This therefore makes an excellent vehicle for both viral genome and viral coat protein replication via the enzyme RNA-dependent RNA polymerase (RdRps) which binds to the 3′ end of the replication template, or open reading frame (ORF). The presence of the uncoated viral genome is recognized by host translation and ribosomal subunits, initiating replication. In brief, the viral genome effectively acts as mRNA, thus providing the codes for synthesis of the enzymes that will in turn initiate the synthesis of new viral genomes. Here again we see how the virus has commandeered normal host functionality.

The modified host genes now replicate viral components that spontaneously begin to reassemble; this process is known as **maturation**. When these daughter virus particles are fully matured, the final stage is their **release** from the host cell as fully matured virions, where they are well placed to infect the next cell. This is normally a passive process, as it is not in the pathogen's best interest to destroy the host.

Passage of the newly formed virions from one part of the plant to another occurs locally, from cell to cell via the **plasmodesmata**, and/or via the vascular system. Viruses that are transmitted via the vascular system are easily recognizable, as the damage caused (for example, lesions in the case of the mosaic viruses) tends to be evenly distributed across the tissue surface (for example, in close proximity to the veins in a leaf). However, this type of movement is dependent on the ability of the infecting virus to suppress RNA silencing (a host defense mechanism that is described in Chapter 4).

Although viral pathogens rarely kill the host plant, the process of re-encoding the host's genome seriously disrupts the normal function of the modified genes. This can be identified by a number of diagnostic morphological changes in the affected host, known as **expression**. In Chapter 2 we described how these morphological expressions are commonly used in taxonomic nomenclature—for example, in tobacco mosaic virus (TMV) that causes mosaic-like lesions on the host leaves. Three other common examples are the leaf-roll viruses, the ring-spot viruses, and the color-break viruses.

Color-Break Virus in Tulips

Although it is difficult to find much that is positive about viral infections of plants, some tulip breeders have utilized the color-break viruses (belonging to the potyvirus group) to induce some striking effects in tulip blooms (Figure 3.16). Gardeners should beware, however, that if this same virus gets carried by a vector to their lilies the effect will be quite damaging, and thus far less pleasing to the eye!

3.8 Other Pathogens

As described earlier, many other pathogens have previously been categorized as members of the other groups. One reason for this is the similarity in structure and life cycles. Fungus-like pathogens such as the Ooomycota share similar lifestyle characteristics with the fungi, and likewise the phytoplasmids share similar lifestyle characteristics with the rest of the domain Bacteria.

However, there are a few unusual pathogens that warrant mention. The *Cephaleuros* species are a group of algae, some of which are parasitic on plants. For example, *Cepahleuros virescens*, commonly known as algal leaf spot or red rust (not to be confused with the fungal rusts), infects mainly the leaves of tropical and subtropical higher plants, including many commercially important species, such as tea, coffee, coconut, mango, guava, and citrus (Figure 3.17). The amount of damage that is caused is dependent on a number of factors, including the host species, host health, and environmental conditions. This pathogen is a filamentous alga that requires a film of water to complete its life cycle. It is therefore typically common in the warm and permanently wet conditions that are found in much of the tropics. Transmitted from host to host by wind or water splash, dwarf sporophytes produce zoospores that penetrate the leaf cuticle, often producing haustorial cells on or under the lower and upper epidermal cells, where they germinate and form disc-like orange-red algal blooms on the leaf surface in

Figure 3.16 Tulip bloom showing the characteristic morphology of color-break virus. A color version of this figure can be found in the color plate section at the end of the book.

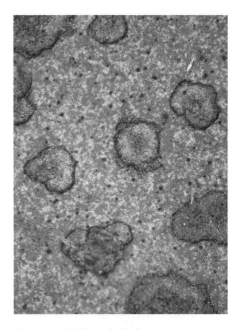

Figure 3.17 The algal plant parasite *Cephaleuros virescens* on leaf of *Citrus* species. A color version of this figure can be found in the color plate section at the end of the book. (Courtesy of Cesar Calderon, USDA APHIS PPQ. Bugwood.org.)

Figure 3.18 **The hemiparasitic plant mistletoe (*Viscum album*).**

the form of thalli with fine filaments. These structures continue to spread asexually, often forming a thallus several centimeters across.

Some *Cephaleuros* species are not always independent pathogens. They are one of the groups of algae that often partner with a fungus to form a symbiotic relationship known as a lichen. A lichen formed with this particular genus of algae became recategorized as *Strigula* species in a quirk of taxonomy whereby a whole new set of taxonomic groupings have superseded the previously described fungal and algal genera and species that have formed their component parts. It is notable that when this alga is part of a lichen, it may or may not still be parasitic.

Many plants are also parasitic on other plants, obtaining all or some of their nutrients from the host. For example, the mistletoes, such as *Viscum album* (Figure 3.18), which commonly grows on many tree species in the UK and parts of mainland Europe, has green leaves and is therefore capable of photosynthesizing to supply some of its nutritional requirements (that is, it is hemiparasitic). However, this plant is an obligate parasite, obtaining its water, minerals, and supplementary photosynthates from the host. Mistletoe is a flowering plant and thus bears seeds. When the seeds germinate they are initially independent of the host, utilizing the nutrients stored in the seed as well as producing sufficient sugars by photosynthesis for the early stages of development, until they are able to form a hemiparasitic relationship with a host.

Many plant parasites have no green tissue at all and are therefore entirely dependent on the host for all of their nutrients (that is, they are holoparasitic). An example of this is the genus *Cuscata* (dodder) (Figure 3.19), which includes many species and has many hosts. It forms dense networks of very fine stems all over the host, has diminutive scale-like, non-photosynthetic leaves, and produces haustoria that penetrate the host vascular tissue to directly obtain the nutrients that it requires. It is a parasitic flowering plant that produces large numbers of tiny seeds that have a long survival time, increasing their opportunities to find new hosts.

Figure 3.19 **The holoparasitic plant dodder (*Cuscata* species).** A color version of this figure can be found in the color plate section at the end of the book.

3.9 Summary

The development of plant pathogens has an extensive history, and evidence of early partnerships (around 400 million years ago) between land plants and fungal organisms provides an insight into the mechanisms of these relationships, many of which show similarities to extant biotrophic infection processes. The development of fungal and bacterial pathogenesis has subsequently evolved complex mechanisms ranging from the secretion of a simple suite of cell-wall-degrading enzymes by the pathogen to an array of complex infection mechanisms as plant pathogens develop new strategies to overcome host defense mechanisms. These new infection strategies rely on chemical signals (elicitors or effector molecules) that mediate the expression of virulence genes that switch on cascade mechanisms, which enhance the infection process. If over time one or more of the virulence genes of the pathogen can be suppressed by the evolution of new resistance genes in the host, the virulence of the pathogen may be reduced. This then leads to a new round of evolution and development in the pathogen. When breeding for disease resistance, plant breeders map and follow the evolution and expression of resistance genes and the emergence of associated race-specific pathogens.

Further Reading

Agrios GN (2005) Parasitism and disease development. In Plant Pathology, 5th ed, pp 77–104. Elsevier Academic Press.

Bancal MO, Hansart A, Sache I & Bancal P (2012) Modelling fungal sink competitiveness with grains for assimilates in wheat infected by a biotrophic pathogen. *Ann Bot* 110:113–123.

Brand T (2009) Powdery mildew of succulent *Euphorbia* species. *Euphorbia World* 5:5–9.

Burdon JJ & Thrall PH (2009) Co-evolution of plants and their pathogens in natural habitats. *Science* 324:755–756.

Büttner D, Noël L, Thieme F & Bonas U (2003) Genomic approaches in *Xanthomonas campestris* pv. *vesicatoria* allow fishing for virulence genes. *J Biotechnol* 106:203–214.

Dodds PN, Lawrence GJ, Catanzariti AM et al. (2006) Direct protein interaction underlies gene-for-gene specificity and coevolution of the flax resistance genes and flax rust avirulence genes. *Proc Natl Acad Sci USA* 103:8888–8893.

Dutton MV & Evans CS (1996) Oxalate production by fungi: its role in pathogenicity and ecology in the soil environment. *Can J Microbiol* 42:881–895.

El Hadrami A, El-Bebany AF, Yao Z et al. (2012) Plants versus fungi and oomycetes: pathogenesis, defense and counter-defense in the proteomics era. *Int J Mol Sci* 13:7237–7259.

Fereres A & Moreno A (2009) Behavioural aspects influencing plant virus transmission by homopteran insects. *Virus Res* 141:158–168.

Horbach R, Navarro-Quesada AR, Knogge W & Deising HB (2011) When and how to kill a plant cell: infection strategies of plant pathogenic fungi. *J Plant Physiol* 168:51–62.

Jones DG (1987) In Plant Pathology: Principles and Practice, Infection & Colonisation pp. 33–43. Open University Press.

Kong L-A, Yang J, Li G-T et al. (2012) Different chitin synthase genes are required for various developmental and plant infection processes in the rice blast fungus *Magnaporthe oryzae*. *PLoS Pathog* 8:e1002526.

Martiniere A, Zancarini A & Drucker M (2009) Aphid transmission of cauliflower mosaic virus. *Plant Signal Behav* 4:548–550.

Meng S, Torto-Alalibo T, Chibucos MC et al. (2009) Common processes in pathogenesis by fungal and oomycete plant pathogens, described with Gene Ontology terms. *BMC Microbiol* 9(Suppl. 1):S7.

Moreno A, Garzo E, Farnandez-Mata G et al. (2011) Aphids secrete watery saliva into plant tissues from the onset of stylet penetration. *Entomol Exp Appl* 139:145–153.

Morris CE, Bardin M, Kinkel LL et al. (2009) Expanding the paradigms of plant pathogen life history and evolution of parasitic fitness beyond agricultural boundaries. *PLoS Pathog* 5:e1000693.

Nelson SC (2008) *Cephaleuros* Species, the Plant-Parasitic Green Algae. Cooperative Extension Service, College of Tropical Agriculture and Human Resources, University of Hawaii. www.ctahr.hawaii.edu/oc/freepubs/pdf/pd-43.pdf

O'Brien D (2012) Protein biomarkers identify disease-carrying aphids. *Agric Res Mag* 60:12.

Oliver RP & Ipcho SVS (2004) *Arabidopsis* pathology breathes new life into the necrotrophs-vs.-biotrophs classification of fungal pathogens. *Mol Plant Pathol* 5:347–352.

Perrine-Walker FM, Prayitno J, Rolfe BG et al. (2007) Infection process and the interaction of rice roots with rhizobia. *J Exp Bot* 58:3343–3350.

Rep M (2005) Small proteins of plant-pathogenic fungi secreted during host colonization. *Microbiol Lett* 253:19–27.

Soto MJ, Sanjuán J & Olivares J (2006) Rhizobia and plant-pathogenic bacteria: common infection weapons. *Microbiology* 152:3167–3174.

Stes E, Vandeputte OM, El Jaziri M et al. (2011) A successful bacterial coup d'état: How *Rhodococcus fascians* redirects plant development. *Annu Rev Phytopathol* 49:69–86.

Taylor TN, Remy W, Hass H & Kerp H (1995) Fossil arbuscular mycorrhizae from the Early Devonian. *Mycologia* 87:560–573.

Tjallingii WF & Hogen Esch TH (1993) Fine structure of aphid stylet routes in plant tissues in correlation with EPG signals. *Physiol Entomol* 18:317–328.

Voegele RT & Mendgen K (2003) Rust haustoria: nutrient uptake and beyond. *New Phytol* 159:93–100.

Webster J & Weber RWS (2007) Introduction to Fungi, 3rd ed. Cambridge University Press.

Xing T, Higgins VJ & Blumwald E (1996) Regulation of plant defence response to fungal pathogens: two types of protein kinases in the reversible phosphorylation of host plasma membrane H^+-ATPase. *Plant Cell* 8:555–564.

Chapter 4

Plant Responses to Pathogens

In Chapter 3 we discussed the processes involved in infection of plants by pathogenic organisms. First, the host and pathogen have to be compatible with each other for infection to occur. Clearly, however, where they are compatible and do interact, if all the odds were on the side of the pathogen, the survival of the host would be at serious risk. Equally, in the longer term, pathogen survival would not be maintained as host availability became increasingly restricted. In order to respond to pathogen attack, plants possess a whole armory of defenses against infection. These defense mechanisms can be physiological (for example, the cuticle, which is a passive permanent physical defense), biochemical (for example, phenols and terpenes that have multiple functions, including antimicrobial properties), and internal (for example, the hypersensitive response that is initiated and/ or up-regulated in response to the presence of pathogens). Basically plants have a programmed system of recognition of and defenses against pathogen infection. Most of the plant host responses are applicable to all pathogen types (that is, fungi, bacteria, viruses, and others). In addition, it should be noted that such responses are directly equivalent to abiotic stress factors. Disease escape, tolerance, and resistance directly equate to **abiotic** stress escape, tolerance, and resistance. Furthermore, recognition of abiotic stress factors activates similar signal transduction pathways and associated gene expression.

4.1 Categories of Plant Defense Against Disease

Disease Escape

Disease escape is based on the principle that prevention is better than cure. There are a number of ways in which plants may escape disease. Susceptibility may be related to a particular growth stage in the host, so the presence of the pathogen at any other stage of plant growth will mean that the plant escapes from disease. Variation in temperature and humidity may also increase or decrease susceptibility if the host and the pathogen have different optimum environmental parameters for development. The distance between potential host plants may render the pathogen unable to infect new hosts. Specifically, if the plants in the zone between two populations of potential host plants are themselves not susceptible, transmission of the pathogen across this zone to the next potential hosts is unlikely to occur. Indeed, the plants in this interim zone may even have antimicrobial or pest-repellent properties. This characteristic is often exploited by growers when they use companion planting. For example, *Allium* species such as onion and garlic, which produce sulfur-based biochemical compounds that repel pests and microbes, can be grown in association with more disease-susceptible crops, such as carrots or brassicas, offering passive resistance to attack by virtue of their proximity. This will be discussed in more detail in Chapter 11.

The plant's epidermal layer forms a protective boundary of cells that covers all parts of the plant (woody stems have a periderm layer in place of the epidermis), and acts as a physical barrier both to environmental challenges and to disease attack. Consequently, even when a pathogen is present the plant may escape infection, due to the presence of its epidermal layer. In most plants the epidermis consists of a single layer of cells comprising a suite of cell types that play specific roles in the protection and maintenance of the plant. Most of the epidermis is composed of cells that are sometimes termed pavement cells due to their "crazy-paving" appearance. These cells are tightly packed both for protection and to provide mechanical strength, and are covered by a cuticle layer made of **cutin** that may have a waxy surface to provide extra protection against water loss in drier environments, but also serves as an additional layer of protection against infection (Figure 3.1). Other cell types in the epidermis include trichomes, which are hair-like structures that help to reduce water loss and sometimes also herbivory in aerial parts of the plant, or that are specialized for the absorption of water and minerals in the roots (where they are generally termed root hairs). The epidermis also contains stomatal guard cells and their subsidiary cells that form the stomatal complex. These are found predominantly on the underside of the leaves in most higher plants. This cell complex regulates the size of the stomata, which are the pores that open and close to facilitate gas and water vapor exchange between the internal parts of the plant and the surrounding atmosphere. (In aquatic plants with floating leaves the stomatal complex is found on the upper surface, to avoid water ingress during gaseous exchange.) Unfortunately, these pores also often function as one of the routes of entry into the plant for pathogens, and especially for fungal hyphae. Pathogens may also enter via wounds in the plant tissues, often incurred by grazing or trampling. The physiology of the plant's defensive layers is described in more detail in Chapter 3.

Disease Tolerance

The concept of co-evolution of plant hosts and their pathogens assumes that pathogen challenge selects for host fitness, and that host resistance selects for pathogen virulence, in a co-evolutionary "arms race." However, in an intermediate position between the pathogen completely overcoming the plant host and the plant host completely resisting the pathogen lies tolerance. In this situation, after disease infection the pathogen makes sufficient gains to sustain itself, and the host suffers only minimal harm, so there is little pressure to up-regulate resistance and counter-resistance in the host and the pathogen, respectively. For the grower, reductions in harvest may not even be noticeable, as they may fall within the normal variation expected for crop yields. One study on infection of *Arabidopsis thaliana* by *Pseudomonas syringae* indicated that, although on average infected plants showed reduced fitness compared with non-infected plants, there was no specific correlation between bacterial growth and symptoms in the host plants, indicating significant variation in tolerance vs. resistance levels (Kover & Schaal, 2002). The mechanism underlying tolerance is not fully understood, but is generally considered to be a host-driven capacity to reduce the pathogen's virulence to a tolerable level.

Disease Resistance

This is described in the next section.

4.2 Disease Resistance

There are six stages of disease resistance:

1. Innate immunity
2. Basal resistance in the plant cells
3. Pathogen counter-responses
4. The hypersensitive response
5. Systemic acquired resistance
6. The "arms race."

Each of these stages will now be described.

The first line of defense is **innate immunity**. This immune response is initiated as plant cells detect and recognize certain proteins, sugars, and other microbial cell wall components, such as peptidoglycan in bacteria, which are collectively termed pathogen-associated molecular patterns (PAMPs) or microbe-associated molecular patterns (MAMPs). Many plants also have specific receptors on their cell surfaces that recognize the presence of pathogens, known as pathogen recognition receptors (PRRs). The innate immunity response tends to be a general response to the presence of microbes, rather than being specific to particular pathogenic organisms, although it does trigger the second line of defense, namely **basal resistance in the plant cells**. This is both physical (for example, reduction in stomatal pore size and thickening of the cuticle) and biochemical (for example, the presence of **phytoanticipins**). Phytoanticipins are a group of multifunctional secondary metabolites that are present as a constitutive part of a plant host's defenses, and are therefore part of the innate immunity response. Table 4.1 shows a representative sample of the phytoanticipins. Other similar compounds are produced only in response to pathogen infection; these are termed **phytoalexins**. However, in each case these compounds fall into a number of biochemical groups, including the phenols and terpenes, and are often referred to as plant antibiotics.

The phytoanticipins function in a number of ways (for example, by interfering with microbial membrane activity, or sometimes by adversely affecting virus vectors). As they are generally significantly volatile, the host plant has to have a number of ways to avoid being damaged by its own defensive compounds. The phytoanticipins may be released into the atmosphere

Table 4.1 Examples of biochemical plant defenses against pathogen attack.

Biochemical Compound	Example	Description	Pathogen(s) Against which Protection is Given
Terpenes	Saponin	Soapy glycosidic compounds produced by most plants	Fungi
	Pyrethrin	Insect neurotoxin produced by chrysanthemums	Virus vectors (for example, aphids)
Phenols	Camalexin	Antibiotic produced by *Arabidopsis* species	Bacteria and fungi
	Salicylic acid	Found in many plants, especially willow trees; works by activating genes that encode anti-pathogenic complexes	Bacteria and fungi
Tannins	—	Flavonoid polymers produced by most plants	Virus vectors (for example, aphids)
Alkaloids	Caffeine	Found in many plants, especially coffee	Fungi and virus vectors
Proteins	Enzymes	Many types, found in all plants; various mechanisms	All pathogens
	Defensins	Cysteine-rich proteins produced by most plants	Bacteria and fungi
Lipid-based compounds	Jasmonates	Made from fatty acids in chloroplast membranes in all plants	Bacteria, fungi, and virus vectors; also play a role in systemic resistance to pathogens in plants

surrounding the host, thus acting as a repellent, or they can be accumulated in dead cells, or in many cases (for example, the **saponins**) they are stored in an inactive form in vacuoles within living cells.

The continuing success of plant pathogens through evolutionary history clearly indicates that they are not without their own system of **pathogen counter-responses** to these defenses. For example, many of the Gram-negative bacteria, such as *Pseudomonas*, *Xanthomonas*, and *Erwinia*, have evolved a system known as a **type III secretion system (TTSS)** that produces an extensive range of effector proteins encoded by *hrp* genes. Other bacteria, fungi, and oomycota have similar systems designed to compromise the plant's defenses in a number of ways—for example, by suppressing plant host enzymes, damaging cell membranes, interfering with host signaling mechanisms, or producing plant hormone-mimicking compounds. A study reported in 2006 (Abramovitch et al., 2006) indicates that plants may down-regulate their auxin production as a defense against pathogen attack, as this induces activation of auxin response factors (ARF). However, some strains of *Pseudomonas* and *Xanthomonas* bacteria can mimic this hormone as a counter-defense to prevent the activation of ARF. Furthermore, some necrotrophic fungal pathogens may impair auxin signaling gene mutants that would normally play a defensive role.

Auxins are not the only plant hormones involved in plant response mechanisms. Recent studies indicate that such defenses are in part facilitated by complex interactions between a number of plant hormone systems, including jasmonic acid, salicylic acid, abscisic acid, and ethylene. The extent of these interactions is currently being characterized, so students with a particular interest in this topic should keep up to date with current research.

The particular suite of effector proteins possessed by any given pathogen is directly related to its virulence. However, in addition to effector proteins, pathogens also produce other virulence factors, such as toxins, that kill off plant cells. These effector molecules interfere with the host plant's biological functions via two different mechanisms, namely cytoplasmic effectors that interact with the host cell organelles (including the nucleus), and apoplastic effectors that interact with host cell surface receptors. **Exopolysaccharides** (EPS) that may block xylem vessels causing wilt symptoms can be equally effective. EPS are long chains of repeating sugar units attached to a carrier lipid and generally complexed with other lipids, proteins, metal ions, organic or inorganic compounds, or DNA. They are sticky compounds secreted by bacteria for a variety of purposes, such as adherence, biofilm production, or, in this case, plant pathogenesis. Clearly, the more means by which the pathogen can affect the host, the more host resources are required for defense against attack.

The evidence for the co-evolution of plants and their associated pathogens is overwhelming, so to these pathogen counter-responses the plant has its own counter-counter-response, known as the **hypersensitive response (HR)**. This is a rapid response activated by the *hrp* gene cluster (hrp stands for hypersensitive response and pathogenicity). This is much more specific than the innate immunity responses in that it is triggered by the presence of pathogen effector proteins (Figure 4.1). When infection is detected, the host plant destroys its own cells immediately surrounding the invading organism, in a process known as programmed cell death (PCD), which is initiated by a number of reactions, including phytoalexin production and an oxidative burst. Under normal circumstances, plants regulate their internal biochemistry by utilizing enzymes and redox metabolites synchronistically. This includes detoxification of reactive oxygen species (ROS) that would otherwise damage plant cells. ROS are formed either as a result of incomplete reduction of oxygen during redox reactions, or by oxidation of

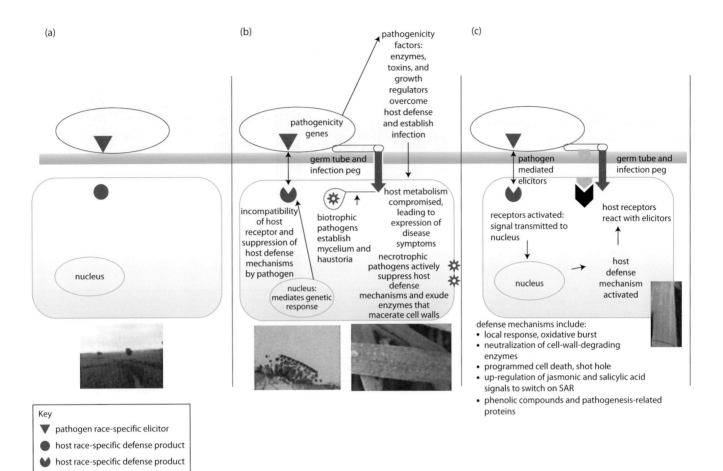

Figure 4.1 Schematic diagram of infection and detection and host response. (a) There is no recognition reaction, so the pathogen does not recognize the host species, and there is no infection. (b) There is a compatible reaction between the host species and the pathogen, leading to germination of the pathogen spore, suppression of the host's defense mechanisms, and successful infection. (c) There is a recognition reaction in which the host has a suite of defense mechanisms to suppress infection by the pathogen. The pathogen may germinate, but successful establishment and subsequent infection are suppressed by the host. SAR, systemic acquired resistance. A color version of this figure can be found in the color plate section at the end of the book.

water by metabolic electron transfer chains in plant cells. However, during the hypersensitive response, ROS are allowed to develop in a reactive burst that contributes to the programmed destruction of cells surrounding the site of pathogen incursion. PCD effectively leaves the pathogen with nowhere to go, thus preventing spread of the disease from that point. In addition, the first row of living cells surrounding the infection site develop a temporary reduction in membrane permeability, as well as structural reinforcement, providing further protection against pathogen ingress. The hypersensitive response is very easily to identify visually, as it appears as spots of discolored tissue (a lesion) surrounding the infection site. Once the pathogen has been killed, the infection site may develop into a "shot hole" (Figure 4.2). It has been postulated that ROS may even directly damage the pathogen itself.

Once the hypersensitive response has been initiated it persists for some time. Research on various plants, including *Arabidopsis*, *Lupinus*, *Nicotiana*, and *Asparagus*, implies that the timescale varies according to the species, but 20 days is not uncommon. The hypersensitive response functions by preventing further attack by the pathogen until it is no longer a danger to the plant. This is termed **systemic acquired resistance (SAR)** (Figure 4.3), and it offers the plant protection from a whole range of pathogens, not just the one that triggered the mechanism in the first place. SAR is in part responsible for

Figure 4.2 Primula leaf with shot hole. This is the result of the hypersensitive response in the plant following infection by the fungal pathogen *Alternaria*. Cells surrounding the infection site are destroyed by the host using programmed cell death, effectively leaving the pathogen with nowhere to go.

Figure 4.3 Schematic diagram of systemic acquired resistance. R, recognition (for example, gene for gene); EL, elicitors; OG, oligogalacturonides; HR, hypersensitive response; RO, reactive oxygen; PAL, phytoalexins; SAR, systemic acquired resistance. A color version of this figure can be found in the color plate section at the end of the book.

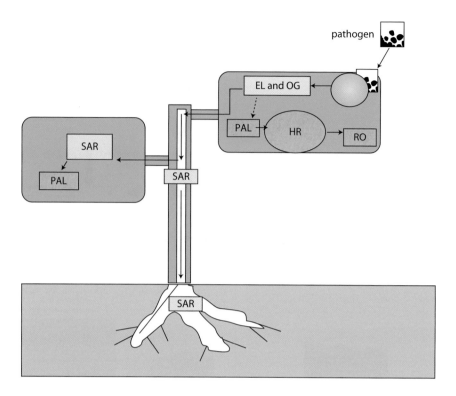

the up-regulation of salicylic acid (one of the phenols, and incidentally the main component of aspirin). Salicylic acid both activates a gene construct known as NPR1 (non-expressor of pathogenesis-related genes 1) and triggers its translocation into the nucleus of the cell. NPR1 then interacts with further transcription factors to mediate downstream expression of full SAR, which is thought to include anti-pathogen enzymes that catalyze the breakdown of pathogen cell-wall components. NPR1 is known to be involved in the regulation of the protein secretory pathway genes that mediate SAR proteins such as the pathogenesis-related (PR) proteins. One such protein, PR1, is commonly reported to be highly expressed and subsequently correlated with enhanced disease resistance. Further evidence for the SAR and NPR1 regulation systems comes from studies using *Arabidopsis* mutants (*secb1a bip2* and *dad1 bip2*) in which the full protein secretory pathway has been impaired. In these mutants the expression of PR1 was significantly reduced and plants were more susceptible to *Pseudomonas syringae* pv. *maculicola* as determined by the number of colony-forming units on infected mutants compared with the wild type.

What is particularly interesting about SAR is that this genetic up-regulation to a primed state of resistance appears to induce a kind of memory within the plant, analogous with animal immunoglobulin resistance. Recent studies have demonstrated that this memory can be passed on to the next generation. For example, in tobacco (*Nicotiana*), parent plants infected with tobacco mosaic virus (TMV) generated progeny that had increased homologous recombination as a result of the TMV infection in the parents, and subsequently the progeny expressed enhanced disease resistance. This phenomenon is known as next-generation systemic acquired resistance (NGSAR), and it is not limited to tobacco. Slaughter et al. (2012) reported that the progeny of *Arabidopsis thaliana* plants treated with SAR elicitors (in this case an avirulent isolate of *Hyaloperonospora arabidopsis*, the causative agent of downy mildew in *A. thaliana*) exhibited enhanced resistance to the disease through the NGSAR mechanism. This resistance does not have long-term durability, as the SAR response disappears in subsequent generations if the antagonist is withdrawn. However, Luna et al. (2012) demonstrated that trans-generational SAR can be maintained for numerous generations if the original antagonist (in this case *Pseudomonas syringae* pv. *Tomato*) persists in the culture environment of the plant, and thus continues to challenge successive generations. Basically plants are capable of adjusting their trans-generational resistance according to the severity of the disease in their growing environment.

Until recently it was considered likely that these sophisticated immune responses were limited to higher plants. However, more recent studies have demonstrated SAR in lower plants, too. In 2014, Winter et al. published their findings on SAR in a moss, *Amblystegium serpens*, infected by the oomycete *Pythium irregular*, suggesting that this type of pathogen resistance evolved prior to the divergence of vascular and non-vascular plants.

Due to the longevity of the effects of SAR, growers are increasingly using environmentally benign chemicals termed plant activators (for example, liquid seaweed extract) that trigger this response, in order to protect their crops. Sometimes, however, all of these defenses are not enough and the pathogen develops the means to overwhelm the plant's defensive systems in a continually evolving **arms race**.

4.3 Gene-for-Gene Theory

It was in the 1940s that the plant pathologist Harold Henry Flor postulated his gene-for-gene theory using rust (*Melampsora lini*) as the typical pathogen and flax (*Linum usitatissimum*) as the typical host. According to Flor, the pathogen and the host have a matching pair of alleles. The pathogen has genes that encode the potential to infect the host—the avirulence (*Avr*) genes. The hosts have matching genes that encode resistance to the pathogen—the resistance (*R*) genes (see Chapter 3). Thus the pathogen and the host have clearly evolved alongside one another. As the invading microbe evolves greater infectivity, so the host evolves greater resistance, each counter-responding to the other, with the time lag between these responses creating a kind of **boom and bust** scenario (**Figure 4.4**).

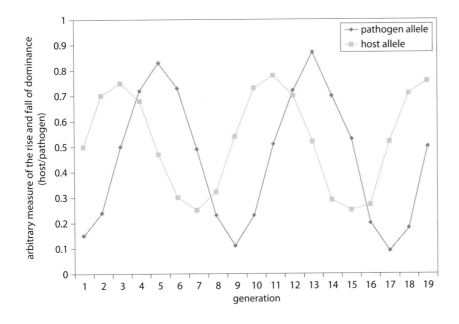

4.1 Dutch Elm Disease: why so vulnerable?

Nobody born after about 1965 will remember the distinct profile of the British landscape before the advent of Dutch elm disease (caused by the fungal pathogen *Ceratocystis ulmi*), which caused an almost complete loss of this one tree genus (*Ulmus*). To put this major change into perspective, imagine the countryside now if all of the oak trees, for example, were suddenly not there anymore. You may not think about them all that much, but you would miss them if they were gone! This disease was so devastating that it swept across the globe, affecting elms in Europe, Asia, and America. It was named "Dutch" elm disease because it was first identified in Holland in the 1920s, although it probably did not originate there. So why is the elm so vulnerable to this disease?

Throughout the chapters of this book so far we have discussed the intimate relationship between host and pathogen, the gene-for-gene concept, and the co-evolution of alleles of these partners. However, the elm could not evolve in this way as its main method of reproduction is via **suckering**, which is a form of vegetative reproduction, so genetically almost all elm trees were identical and therefore unchanging. Once the pathogen took hold, the host simply could not keep pace with it. Furthermore, the fungus is carried by a bark beetle that typically infests elm trees as they mature, reducing the possibility of seed production, and thus the chances of genetic variation. As if this was not problematic enough, when elms do produce seeds they are often sterile. However, there are a few elms left in the UK, the largest collection being in Brighton. There the elms are constantly monitored for disease, and any fertile seeds that are produced are collected with the long-term aim of replanting. First, however, this disease needs to be tackled!

Figure 4.4 Graph illustrating the principle of boom and bust in relation to the gene-for-gene hypothesis regarding the co-evolution of plant hosts and their respective pathogens.

4.4 Suicide Genes?

For a long time *Avr* genes appeared to be almost self-defeating. In themselves, and by definition not virulent and therefore not disease inducing, these genes are largely responsible for eliciting plant host defenses. Clearly, however, pathogens would not have evolved and retained genes whose sole purpose is to activate counter-responses against them. Avirulence in pathogens does have the advantage of allowing infection of the host "through the back door", as plant responses to pathogens that are not expressing virulence are less extreme. However, the *Avr* genes do in fact encode a suite of proteins that interact with the pathogen's virulence genes, thus aiding and promoting infection and disease in the host. The exact nature of these interactions has yet to be defined, and is the subject of a significant amount of ongoing research.

4.5 Endophytes

Infection of plants by fungi and bacteria does not always lead directly to disease. Indeed many such microorganisms live symbiotically within the plant as **endophytes**, (from the Greek *endo*, meaning "inside," and *phyte*, meaning "plant"), causing no detrimental effects at all, and in most cases this symbiosis is mutualistic—that is, each partner provides one or more benefits to the other. Generally the endophyte gains an environment in which it can thrive, a means of reproductive dispersal, and nutrients in the form of photosynthates from the plant host, and the plant gains an increased capacity for uptake of mineral nutrients via the endophyte. Endophytes infect their host plants in the same ways as do pathogenic fungi and bacteria, and indeed most endophytes have close relatives among the pathogens. For instance, the genus *Verticillium* comprises mainly pathogenic species that cause wilt diseases in many plants, but a few species exist specifically as endophytes. Like the pathogens, endophytes may be generalists, infecting many plant types, or specialists, infecting a narrow range of hosts. By the same token, the plant host may have only a few or a wide range of endophytes associated with it. To date, no plants have been found without any at all!

There is a growing focus of research on the additional role of many endophytes in preventing incursion by and/or defending against infection by pathogenic organisms, either by competition or by the production of secondary metabolites such as antibiotics (Table 4.2). Indeed, many such endophytes offer resistance to pathogens belonging to their own genus. For example, Tyvaert et al. (2014) published evidence that the *Verticillium* Vt305 endophyte species reduced symptoms and tissue colonization of *Verticillium longisporum* (a wilt-inducing species) in cauliflower plants.

Table 4.2 Examples of antibiotics produced by fungal endophytes, the associated plant hosts, and target pathogens.

Fungal Endophyte	Host Plant (Common Name)	Host Plant (Latin Name)	Antibiotic	Target Pathogens
Acremonium zeae	Maize	*Zea mays*	Pyrrocidines A and B	*Aspergillus flavus, Fusarium verticillioides*
Verticillium species	A Chinese medicinal herb	*Rehmannia glutinosa*	Massariphenone, ergosterol peroxide	*Pyricularia oryzae* P-2b
Phomopsis cassiae	Cassia tree	*Cassia spectabilis*	Cadinene sesquiterpenes	*Cladosporium sphaerospermum, C. cladosporioides*
Periconia species	Japanese yew tree	*Taxus cuspidata*	Fusicoccane diterpenes	*Bacillus subtilis, Staphylococcus aureus, Klebsiella pneumoniae, Salmonella typhimurium*
Ampelomyces species	Prickly goldenfleece	*Urospermum picroides*	3-O-methylalaternin, altersolanol A	*Staphylococcus aureus, S. epidermidis, Enterococcus faecalis*

Adapted from Gao et al. (2010).

The endophytes also increase the host's resistance to abiotic stresses such as drought, and help to reinforce resistance to herbivory. There is of course a degree of self-interest involved, as the endophyte would otherwise have to compete with pathogens for the host's photosynthetic products. It would also be detrimental to the endophyte if its plant host was in poor health due to abiotic stresses and/or herbivory, and therefore less well equipped to provide the excess nutrients required for the survival of the endophyte.

One example of this relationship is the interaction between endosymbiotic fungi of the genus *Neotyphodium* (an obligate symbiont) and a number of temperate perennial grasses, including the fescues (*Festuca* species*)* and the ryegrasses (*Lolium* species). *Neotyphodium* performs a number of useful functions that benefit the grass. As well as up-regulating nutrient acquisition, it outcompetes other fungi, thereby preventing pathogen ingress, it produces alkaloids that are unpalatable to herbivorous pests, and it enhances plant mechanisms that protect against abiotic stresses. Indeed, recent research suggests that the endosymbiont produces ROS that in turn induce the plant to produce antioxidants to protect it against disease, drought, heavy metals, and other oxidative stressors.

It should be noted that some infecting microorganisms have a dual role, as either a pathogen or a beneficial endophyte, often in part depending on the genotype of the host. In a recent study, isolates of pathogenic *Verticillium dahliae* from eggplants were shown to be endophytic in tomato plants, and indeed protected the host plant from infection by virulent *V. dahlia* (Shittu et al., 2009).

4.6 Endophytes and Farmers

For farmers, endophytes often have a downside. For example, endophytes that are associated with grasses used for grazing, such as the ryegrasses (*Lolium* species) and the fescues (*Festuca* species), may keep the host plants healthy, but production of potentially toxic secondary metabolites (for example, alkaloids) that accumulate in the leaf sheaths of the grass can cause diseases in livestock. For instance, perennial ryegrass toxicosis (otherwise known as staggers, and which should not be confused with annual ryegrass toxicity) is a debilitating disease that affects the animal's ability to walk properly, and raises body temperature. Affected animals often try to cool off in water or mud, but once they are down they struggle to get up, and serious injury or even drowning may occur. Staggers can also reduce milk production and affect growth and development of the offspring of livestock.

Furthermore, the extra vigor conferred on the grass species in a mixed grazing sward may allow them to outcompete other species, such as clover (*Trifolium* species) and other legumes from within the original seed mix, thereby reducing the nutritional value of the grassland for the stock animals.

4.7 Plant Host Cell Responses to Viral Infection

In this chapter we have discussed plant host responses to pathogen attack in general terms. These responses are applicable across the range of potential pathogens (for example, fungi, bacteria, and oomycetes). This is largely because the fundamental structure and functionality of all living organisms have significant commonality. That is, they have a carbon-based cellular structure, they need water for survival, they require nutrients for growth, development, and reproduction, and they manufacture similar biochemical compounds (for example, proteins, nucleic acids, polysaccharides, and lipids) from acquired components. In short, at their most basic level all living organisms are built in the same way and have the same needs, whether they are a single-celled microbe or a multicellular plant or animal. These host responses do also apply to a greater or lesser extent to defense against viruses. However, in most respects viruses are somewhat different. Although some of their structural components are similar (that is, they are composed of genetic material and proteins), their lifestyle requires them to be hitch-hikers, extraneous to the life flow of their hosts, on whom they are entirely dependent (see Chapter 2). Plants therefore require extra measures over and above their standard host responses to overcome or manage infection by viruses.

Plant virus infections do not usually kill the host, but they can cause major disruptions to biological processes, and in the case of commercial crops they often significantly reduce productivity. In recent years, plant scientists and molecular biologists have been carrying out studies aimed at identifying infection processes, plant immune responses, and viral counter-responses. This research is essential because, as the world population continues to increase at an alarming rate, any crop yield losses are significant. Thus plant viruses are now regarded as major players in the global battle for food security. Of particular concern are those viruses that endanger production of staple crops, such as rice, potatoes, and cassava (Research Box 4.1).

RESEARCH BOX 4.1 CASSAVA BROWN STREAK DISEASE

The root of the plant cassava (*Manihot esculenta*) (Figure 1), which is a member of the family Euphorbiaceae, is one of the major carbohydrate sources in tropical and some subtropical regions globally, especially Africa, Asia, and also South America, from where it originates. Around 11% of the world's population use cassava as part of their staple diet. The plant's use in ethanol production is also gaining prominence in many developing countries. Thus its economic importance cannot be overstated. More than 50% of the world's cassava is produced in the sub-Saharan African states, and consequently diseases that affect African cassava crops present a significant threat to food security on that continent.

Cassava is affected by a whole range of pathogens, including numerous bacterial, fungal, and other diseases, such as those caused by *Xanthomonas*, *Fusarium*, and *Pythium* genera, respectively. However, one of the diseases of most concern currently is cassava brown streak disease (CBSD) (Figure 2), which is caused by two potyviruses, namely cassava brown

streak virus (CBSV), which belongs to the genus *Ipomovirus*, and the related species Ugandan cassava brown streak virus (UCBSV), both of which are common in Eastern Africa. At the time of writing, these viruses appear to be almost specific to cassava; the only other possible host that has been reported to date is the related *Manihot glaziovii* (Mbanzibwa et al., 2011). Yields of crops infected by this disease can be reduced by up to 50%, with devastating effects on many subsistence farming economies. Although symptoms do appear on the leaves and stems in extreme cases, the edible root is the most severely affected part of the plant, and in cases where the aerial tissues appear to be unaffected the farmer may not realize that there is a major disease incursion until harvest time.

CBSD is transmitted by the silverleaf whitefly (*Bemisia tabaci*), which is also known to be a vector of cassava mosaic virus (CMV), a well-studied and widespread gemini virus. However, significant disease outbreaks have also resulted from vegetative propagation, so newer techniques using improved

Figure 1 Cassava plant (*Manihot esculenta*).

Figure 2 Roots of the cassava plant (*Manihot esculenta*) showing distinctive brown discoloration resulting from infection by cassava brown streak virus (CBSV). Courtesy of Catherine Njuguna, International Institute of Tropical Agriculture (IITA).

RESEARCH BOX 4.1 CASSAVA BROWN STREAK DISEASE

sanitation measures during propagation, together with the use of tissue culture methods, are currently being promoted by agriculturalists to tackle this disease.

Unlike CMV, CBSD has been poorly studied until recently. It appeared to affect only small areas of Eastern Africa, and only at an altitude of over 1000 meters above sea level, and was generally tackled by adopting improved phytosanitation measures. Recently, however, the disease has become more virulent, reaching epidemic proportions in some areas. It is no longer limited by altitude, and is now spreading widely across both Eastern and Central Africa. As a result, the virus now poses one of the most significant threats to cassava yields, and is therefore attracting the attention of agronomists and researchers.

Some research workers have been looking at potentially tackling the whitefly vector populations. However, this involves the widespread use of expensive pesticides. On large-scale cropping systems this approach may be effective as a short-term solution, but most cassava producers are subsistence farmers and smallholders who are unable to afford such treatments. Furthermore, the areas where insecticides are not used are safe havens for the rapidly reproducing whitefly populations. Consequently there is a need for cheaper and more sustainable methods of control, such as biological control and the breeding of resistant varieties of cassava (Legg et al., 2014).

Cross-breeding of the three main varieties—"Albert", "Kiroba", and "Kaleso"—may confer some resistance, as the "Kaleso" variety already possesses some resistant genes. However, the level of success achieved by traditional plant breeding methods can be more variable, so identification of the resistant genes and modification of plants for commercial use is likely to be a more effective approach, and is the focus of ongoing research. This is proving to be a complicated endeavor, as currently there appears to be little correlation between viral load and symptom expression (Kaweesi et al., 2014). In addition, CBSV appears to be adversely affected by the presence of other viral antagonists. All of these factors decrease the reliability of the data.

At the time of writing, the most promising research is focusing on the use of molecular methods to develop resistance in the host plants. In 2011, Patil et al. published research which indicated that high levels of resistance (generally around 85%, but up to 100% in some sample isolates) were promoted using RNA interference (RNAi) technology on highly conserved regions of viral coat proteins.

Research into this very recent threat to cassava crops is still in its infancy, and students with an interest in the topic should keep up to date with current work on this emerging problem.

Once a virus has infected the plant host, resistance is all about inhibition. The hypersensitive response is a strategy used against viruses as well as other pathogens (see above). After viral infection has been detected, the host plant may kill the cells surrounding the infected cells, so that new virus particles formed by replication of the virus are unable to travel the distance to the adjacent living cells. This strategy may work where infection is limited to isolated areas of host tissue. The plant may produce barriers to cell-to-cell movement of the virus by reinforcing the cellular structure, facilitated by a partial rearrangement of the cellular cytoskeleton, cross-linking between glycoproteins and polysaccharides in the cell wall, and up-regulation of lignin polymer synthesis. However, viral infections are commonly transmitted via the host plant's vascular system with the aid of vectors (Research Box 3.1), giving the virus access to all cells in every part of the plant. Inevitably, plant mechanisms of resistance to viruses, over and above the systems described above, must occur at the molecular level. This is because extracellular viruses are in a dormant state, and viruses only become active within the cell when they are undergoing replication.

There are a number of genetically based means by which plants can resist virus incursion. Within the plant's germplasm, resistance is conferred by two types of genes, namely dominant and recessive resistance genes. Dominant resistance is similar to other types of plant pathogen resistance (described above) in that pathogen avirulence genes may induce a hypersensitive response. Recessive resistance is generally, although not always, virus specific. It is based largely on the encoding of translation initiation factors, in that host plant genes mutate and/or fail to produce some of the components required by the virus for replication. Around 100 (about 50%) of the known virus-resistance genes are recessively inherited.

4.8 RNA Silencing

Another resistance factor involves silencing of genes to prevent their expression. This generally occurs post transcription, and is known as post-transcriptional gene silencing (PTGS). It involves the parent strand of nucleic acid base pairing with messenger RNA (mRNA) and forming an RNA-induced silencing complex (RISC). The cleavage mechanism that separates the parent strand from the newly transcribed strand is triggered by endonuclease and **argonaute proteins** that are also part of the RISC. As a result, the newly transcribed strand, encoded by the viral pathogen's nucleic acid, breaks down and translation is prevented, so that production of the components for manufacture of new virions is incomplete.

4.9 Resistance, Counter-Resistance, and Counter-Counter-Resistance

RNA silencing is a mechanism initiated as a result of virus infection, both at the single-cell level and at the systemic level. However, viruses can encode RNA silencing suppressor proteins to counter this response. By the same token, viruses themselves can be utilized by geneticists as a tool to stimulate the silencing process; this is known as virus-induced gene silencing (VIGS). It appears that in these cases the host and the pathogen utilize each other's resistance mechanisms against one another—and so the arms race continues.

The study of viruses, by definition, falls within the topics of genetics and molecular biology. Although we have briefly touched on this topic here, as students need to be aware of the impacts of viruses on plant pathogenesis, a more detailed discussion is beyond the remit of this text. A vast amount of research on the effects of viral pathology on plants is currently ongoing, and students with a particular interest in this topic will find accounts of such research readily available. Here we simply wish to highlight the fact that plant responses to viral infections are very similar to their responses to other pathogenic organisms, but that additional mechanisms are involved in plant responses to viruses. In terms of our understanding of these mechanisms, research is still in the very early stages.

4.10 Summary

While it is essential for scientists and growers alike to understand the nature of pathogenicity, it is equally important to understand how plant hosts react to pathogen challenges, and in turn the response mechanisms of the pathogen to this. Co-evolution dictates a range of significant responses and counter-responses, as hosts and pathogens are typically closely associated, and learning to recognize such relationships is fundamental to the development of methods for exploiting and enhancing the host's own measures that are in place to counter disease attack. At the same time, introduced crops and other plants may fall foul of disease-inducing pathogens, and sometimes have little or no inbuilt resistance, so our ability to enhance or promote, for example, resistance protein production may well be key to the long-term success of the crop. Furthermore, we can only speculate about the subtlety and diversity of factors that will ensue as a result of global climate change (for example, changes in flowering or fruiting times in host plants, early pathogen development, vector organisms undergoing a population crash or a population explosion, and increased or decreased water availability at critical points in the development of both host and pathogen). What is certain is that more research is needed now and will continue to be required as environmental changes progress.

Further Reading

Abramovitch RB, Anderson JC & Martin GB (2006) Bacterial elicitation and evasion of plant innate immunity. *Nat Rev Mol Cell Biol* 7:601–611.

Agrios GN (2005) Plant Pathology, 5th ed. Elsevier Academic Press.

Alfano JR & Collmer A (2004) Type III secretion system effector proteins: double agents in bacterial disease and plant defence. *Annu Rev Phytopathol* 42:385–414.

Conrath U (2006) Systemic acquired resistance. *Plant Signal Behav* 1:179–184.

De Gara L, de Pinto MC & Tommasi F (2003) The antioxidant systems vis-à-vis reactive oxygen species duting plant–pathogen interaction. *Plant Physiol Biochem* 41:863–870.

Fraile A & Garcia-Arenal F (2010) The coevolution of plants and viruses: resistance and pathogenicity. *Adv Virus Res* 76:1–32.

Freeman BC & Beattie GA (2008) An overview of plant defenses against pathogens and herbivores. *Plant Health Instructor* (doi: 10.1094/PHI-I-2008-0226-01).

Gao FK, Dai CC & Liu XZ (2010) Mechanisms of fungal endophytes in plant protection against pathogens. *Afr J Microbiol Res* 4:1346–1351.

Kaweesi T, Kawuki R, Kyaligonza V et al. (2014) Field evaluation of selected cassava genotypes for cassava brown streak disease based on symptom expression and virus load. *Virol J* 11:216.

Kover PX & Schaal BA (2002) Genetic variation for disease resistance and tolerance among *Arabidopsis thaliana* accessions. *Proc Natl Acad Sci USA* 99:11270–11274.

Legg JP, Shirima R, Tajebe LS et al. (2014) Biology and management of *Bemisia* whitefly vectors of cassava virus pandemics in Africa. *Pest Manag Sci* 70:1446–1453.

Llorents F, Muskett P, Sánchez-Vallet A et al. (2008) Repression of the auxin response pathway increases *Arabidopsis* susceptibility to necrotrophic fungi. *Mol Plant* 1:496–509.

Luna E, Bruce TJA, Roberts MR et al. (2012). Next-generation systemic acquired resistance. *Plant Physiol* 158:844–853.

Mbanzibwa DR, Tian YP, Tugume AK et al. (2011) Simultaneous virus-specific detection of the two cassava brown streak-associated viruses by RT-PCR reveals wide distribution in East Africa, mixed infections, and infections in *Manihot glaziovii. J Virol Methods* 171:394–400.

Navarro L, Dunoyer P, Jay F et al. (2006) A plant miRNA contributes to antibacterial resistance by repressing auxin signaling. *Science* 312:436–439.

Patil BL, Ogwok E, Wagaba H et al. (2011) RNAi-mediated resistance to diverse isolates belonging to two virus species involved in cassava brown streak disease. *Mol Plant Pathol* 12:31–41.

Quigley P & Reed K (1999) Endophyte in Perennial Grasses: Effect on Host Plants and Livestock. http://agriculture.vic.gov.au/agriculture/pests-diseases-and-weeds/plant-diseases/grains-pulses-and-cereals/endophyte-in-perennial-grasses-effect-on-host-plants-and-livestock

Rosenberg E, DeLong EF, Stackebrandt E et al. (eds) (2013) The Prokaryotes: Applied Bacteriology and Biotechnology, 4th ed. Springer-Verlag.

Shittu HO, Shakir AS, Nazar RN & Robb J (2009) Endophyte-induced *Verticillium* protection in tomato is range-restricted. *Plant Signal Behav* 4:160–161.

Slaughter A, Daniel X, Flors V et al. (2012) Descendants of primed Arabidopsis plants exhibit enhanced resistance to biotic stress. *Plant Physiol* 158:835–843.

Subramanian S, Sangha JS, Gray BA et al. (2011) Extracts of the marine brown macroalga, *Ascophyllum nodosum*, induce jasmonic acid dependent systemic resistance in *Arabidopsis thaliana* against *Pseudomonas syringae* pv. tomato DC3000 and *Sclerotinia sclerotiorum. Eur J Plant Pathol* 131:237–248.

Truniger V & Aranda MA (2009) Recessive resistance to plant viruses. *Adv Virus Res* 75:119–159.

Tyvaert L, França SC, Debode J & Höfte M (2014) The endophyte *Verticillium* Vt305 protects cauliflower against Verticillium wilt. *J Appl Microbiol* 116:1563–1571.

Wang D, Weaver ND, Kesarwani M & Dong X (2005) Induction of protein secretory pathway is required for systemic acquired resistance. *Science* 308:1036–1040.

White FF, Yang B & Johnson LB (2000) Prospects for understanding avirulence gene function. *Curr Opin Plant Biol* 3:291–298.

White JF Jr & Torres MS (2009) Is plant endophyte-mediated defensive mutualism the result of oxidative stress protection? *Physiol Plant* 138:440–446.

Winter PS, Bowman CE, Villani PJ et al. (2014) Systemic acquired resistance in moss: further evidence for conserved defense mechanisms in plants. *PLoS One* 9:e101880.

Chapter 5
Epidemiology

Epidemiology is the study of the methods by which diseases spread throughout a population. There are a number of factors that influence epidemiological processes, including climate, proximity of a host to its neighbor, host genetic characteristics and variability within that population, and the presence of relevant vectors. This is a particularly pertinent topic with regard to plant pathology, due to the vast expanses of crop monocultures that exist on our planet. Where such plants are vulnerable to any given disease, there is little to stop the progress of large-scale infection (an epidemic). In addition, the dwindling gene pool of many of our major crop plants, such as wheat, potatoes, and rice, has left growers with few options for selection of alternative varieties.

Under natural conditions, plants—like all living organisms—are constantly being exposed to pathogens. However, plants in natural communities normally escape the debilitating epidemics that are observed in cropping systems. In natural ecosystems the balance of species composition favors complex assemblages of plants, with numerous species making up any given habitat. For example, grasslands are composed of numerous grass species often intermixed with herbaceous plants, creating a diverse mosaic of plant species. Of course there are exceptions, such as climax forest, in which eventually one or two tree species dominate the forest canopy. However, even these climax communities seldom experience disease epidemics, and when epidemics are observed in forests they are often the consequence of human activity. The question that then needs to be addressed is why disease epidemics occur in our cropping systems. A cursory inspection of this problem suggests that the answer is simple—monocultures of single crops or at most just a few crop plants dominate our global landscape. However, even then such epidemics are not common. This is due to a number of reasons.

For a pathogen to develop into an epidemic, a number of key factors are needed to tip the balance of probabilities in its favor, namely the presence of a pathogen, the presence of susceptible hosts, and favorable environmental conditions. Pathologists refer to this set of scenarios as the disease triangle (Figure 5.1), and epidemiologists build models to predict disease by developing knowledge of how these parameters influence the establishment and spread of a given disease throughout a cropping region and season (Table 5.1). This chapter investigates the disease triangle in depth, and explores how knowledge of the components of the disease triangle can be used to build models of disease epidemics.

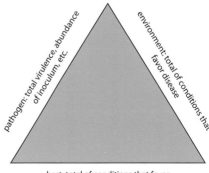

Figure 5.1 Disease triangle. A successful disease cycle needs three components to be aligned: (1) the pathogen must be present; (2) the host must be susceptible; and (3) the environment must be conducive to the life cycle of the pathogen.

Table 5.1 Summary example of key information required when developing a disease model.

Biotic Factors		Examples	Relevance to Modeling	Environmental Factors	Relevance to Modeling
Cropping system	Annual	Annual cereal	Dominated by wheat in the UK, and some regions have second wheats	Temperature	Moderates rate of plant growth and the infection rate of any pathogen. Warmer temperatures favor both plant and pathogen growth
	Perennial	Fruit (soft, cane, and top fruit)	Top fruit (apples and pears) can have significant disease inoculum carryover (for example, sclerotia of brown rot)	Wind	Acts as an agent of dispersal
	Broad leaf	Oilseed rape, cabbages, cauliflower	All of these are brassicas and can be hosts for overwintering inoculum, large stubble, and green bridge. Large and leafy plants increase near-neighbor distance	Leaf wetness and relative humidity	High relative humidity enhances germination rates of pathogen spores such as conidia and urediospores. Periods of leaf wetness can enhance virulence of foliar pathogens
	Root and tuber crop	Potatoes, carrot, swede	Important source of overwintering inoculum. Poor sanitation of seed bed, missed tubers during harvest and cultivation which can act as a potential source of volunteer inoculum. Potatoes have large leafy canopy conditions that favor large potato blight infestations	Light	Variation in light intensity has both positive and negative effects on the biology of plant pathogens. Under good light conditions pathogens may produce reproductive spores, but in low light intensities, such as occur in autumn and winter, pathogens will produce overwintering sclerotia
	Position in rotation	Continuous cropping (for example, wheat → wheat)	This cropping system would favor the development of take-all, a serious stem and root disease in wheat)	pH	Soil chemistry can mediate the presence of plant pathogens. Infections of club root (*Plasmodiophora brassicae*) may be mediated by raising the pH of soils to pH 7 using lime. However, caution must be exercised if potatoes are a component of the rotation, as high soil pH favors common scab (*Streptomyces* species)
Crop growth stage	Cereal growth stage	Zadok decimal growth stage	Depends on initial incidence of disease. Early signs of polycyclic disease at young growth stage can lead to yield loss and increased levels of inoculum in the cropping season. Timing of flag leaf in cereals		
	Early development of crop	Early sown and well established in winter	May be more resistant to disease development and thus check the rate of disease spread		
	Late development of crop	Late sown and poorly developed	Crop is physiologically disadvantaged and thus vulnerable to infection and rapid spread of disease		
Crop genetics	Gene-for-gene theory	Avirulence genes in flax rust (*Melampsora lini*), AVrM, AVrP4, and AVrP123 induce the hypersensitive response in flax due to the presence of the corresponding resistance genes *M*, *P1*, *P2*, and *P4* in flax	Any resistance reaction by the host will reduce the rate of infection and hence progress of disease in the crop		

5.1 Components of Epidemics

Impact of Environmental Factors on Pathogens

Climate is geographically variable, and under normal conditions plants have evolved many mechanisms to enable them to cope within a given climatic region. However, human activity has led to the conversion of many complex species-rich habitats into large areas of uniform environments dominated by a few common crop species. This single act of habitat modification has had a profound impact on the climate of a given cropping system by modifying climatic factors (for example, temperature, relative humidity, leaf surface wetness, windiness, soil pH) to such an extent that the health of a crop is compromised by the presence of local and migrating pathogens. To consider these factors further it is useful to consider the potential outcome of a disease at the individual plant level, and then to scale this up to a crop system.

When a pathogen challenges a plant there are a number of possible outcomes. First, establishment of a pathogen in or on a host may be successful, in which case plant health is compromised by the pathogen, possibly eventually resulting in complete plant death or collapse of a crop. Second, the host may be able to respond to the pathogen and survive the challenge. Third, the host and the pathogen may establish a mutualistic relationship that benefits the pathogen, the host, or both partners. Such a relationship can often result in a symbiotic partnership, as is seen in many mycorrhizal associations. Scaling up the first of these possible outcomes to the level of a crop epidemic, we first need to consider the impact of the crop environment (that is, temperature, relative humidity, leaf surface wetness, and windiness) on the successful colonization of a plant by a pathogen.

Useful models for these discussions are those based on crop production and ornamental plant production systems, with excellent examples from protected cropping systems and arable crops, which are infected in particular by the rusts and mildews, which are biotrophic pathogens. In order to be successful, both of these groups of parasites require certain climatic conditions to be met, and these are basically the same conditions as those required and therefore promoted by the grower. This concept applies at the regional level, within a cropping system, and at the level of the individual plant and leaf surface. Therefore the very act of growing a modern crop inevitably favors disease, and if a given disease is left unchecked this is likely to result in a regional epidemic.

Temperature

Generally, with increasing temperature there is a concurrent rise in overall biological activity until a given threshold is reached, and pathogens are no exception to this rule. Many pathogens can survive cold winters in a resting state (either as a resting spore or as a resting mycelium), and at the onset of warmer conditions sporulation and growth begin. The mildews start to sporulate at quite modest temperatures. For example, for the conidia of rose powdery mildew (*Sphaerotheca pannosa* var. *rosae*) the minimum germination temperature is 5°C, the maximum sporulation temperature is 35°C, and the optimum sporulation temperature is 22°C. It is of course no coincidence that the rose also has an optimum temperature range that is well within the temperature limits for mildew sporulation temperatures.

A very common disease in native British oaks (*Quercus robur* and *Q. petraea*) is oak powdery mildew (*Erysiphe alphitoides*), and the optimum temperature range for conidia germination is within the range of 20–35°C, with good germ tube elongation at temperatures of 25°C. Germination and germ tube growth are inhibited at temperatures above 30°C, and overall colony growth is optimal within the range of 10– 30°C, with complete cessation of pathogen

Figure 5.2 Oak powdery mildew in a hedgerow oak.

growth at 34°C. Again the temperature range that favors oak powdery mildew is well within the range of optimum leaf growth temperatures for British oak trees. These growth temperatures, combined with the wet humid climate of the UK, mean that oak powdery mildew is widespread and easily observed on the leaves of hedgerow trees from July to October (Figure 5.2).

The sporulation and growth of the cereal rusts respond in a very similar pattern to the mildews described above. The causal agent of cereal rusts, *Puccinia graminis* (there are many formal species of cereal rusts; for further details see Chapter 6), has a complex range of different spore types, namely aeciospores, basidiospores, pycniospores, teliospores, and urediniospores, with each spore type having a specific optimum temperature range for sporulation (Table 5.2).

Relative Humidity

Sporulation of pathogen propagules is also strongly correlated with the prevailing relative humidity (RH), and in many cases pathogen growth is enhanced under conditions of moderate to high RH.

The mildews require relatively humid conditions (23–99% RH) for successful host colonization and propagation. However, spore germination in the powdery mildews is decreased if free water is present on the leaf surface, probably due to localized anoxic conditions. In addition, it has been demonstrated that the water relations of spores from one of the mildews, namely the conidia from *Erysiphe polygoni*, have evolved to utilize water vapor directly. The protoplasts of conidia from *E. polygoni* were observed to shrink only very slightly following plasmolysis, which suggested that they contain minimal free water, and furthermore the osmotic potential of the conidial cell sap was determined to be –6300 kPa for *E. polygoni* and –6800 kPa for *E. graminis* f. sp. *Hordei* (Carroll & Wilcox, 2003). Thus it has been proposed that mildews have a high affinity for water, which they absorb directly from the atmosphere, thereby enhancing their pathogenicity under a range of environmental conditions in which free water may be limited but atmospheric water vapor is readily available. Thus free water on the leaf surface may prevent spore germination in the powdery mildews, but the development of this disease is mediated by high RH. This combination of conditions is commonly observed in protected cropping systems, such as greenhouse salad crops. Moreover, conditions of high RH are common not only in the tropics but also in many other parts of the world, including western regions of the UK. Such prevailing climatic conditions favor widespread epidemics of mildews, particularly in protected cropping systems.

Mildew outbreaks are exceptionally widespread in greenhouses, and cause significant economic losses to growers, in terms of both aesthetics and, in the case of downy mildews, eventual plant death. Good examples of powdery mildews are those that infect tomatoes (*Solanum lycopersicum*), an important greenhouse-grown crop around the world, including in the UK. There are

Table 5.2 Rust spore types and their associated sporulation temperature.

Spore Type	Host Plant	Temperature Range (°C)	Optimum Temperature (°C)
Aeciospore	Alternate host (for example, berberis)	2–26	19–20
Basidiospore	Cereal	15–30	15–20
Pycniospore	Alternate host (for example, berberis)	8–35	19–21
Urediniospore	Cereal	10–30	18–20
Teliospore	Cereal	7–28	25

two causative agents of tomato powdery mildew, namely *Leveillula taurica* and *Erysiphe orontii* (the latter was first reported in the UK in 1987). Further taxonomic studies of *E. orontii* have indicated that it should be classified as a member of the genus *Oidium*, and a global study of the *Oidium* genus has highlighted two species: *Oidium neolycopersici* is found in various locations across the globe, except in Australia, where *O. lycopersici* is common.

O. neolycopersici causes mildew symptoms on all the aerial parts of tomato, including powdery white lesions on the leaf blades, petioles, stems, and sepals, but not on the fruits. Experimental manipulation of the RH of tomato leaves by Jacob et al. (2008) demonstrated the effect of RH on the germination and growth of *O. neolycopersici*. A high rate of appressorium formation was observed when RH was in the range of 33–99%. However, at RH values in the range of 7–23%, appressorium formation was significantly reduced. Similar observations were made for other germination parameters, such as germ-tube length, which for the lower RH range was less than 5 μm, compared with an average of 5.13 ± 2.14 μm in germ tubes of all conidia. Where germination had occurred, germ tube length for the lower RH range was less than 15 μm, compared with an average of 27.4 ± 5.13 μm. The observation that germinating conidia and associated establishment of pathological structures connected with powdery mildew are closely linked to the RH of the cropping environment illustrates one of the challenging parameters associated with the epidemiology and subsequent modeling of plant pathogens. The general cycle of pathogenesis of powdery mildew demonstrates how this pathogen has become a successful parasite in protected cropping situations, due to the prevailing risk of high RH levels in the greenhouse.

These high RH levels not only increase the amount of tissue infected by pathogens, but can also mediate the amount of inoculum present in a crop. Studies that manipulated the RH of the culture environment of grape seedlings, which were simultaneously infected with the causative agent of grape powdery mildew (*Uncinula necator*), revealed that the density of conidia and the length of conidial chains were directly proportional to the humidity of the culture environment (Carroll & Wilcox, 2003). There was a linear relationship with RH, where both the incidence and the severity of the disease increased with increasing RH up to an optimal RH of 85%. These studies were conducted at the optimal temperature of 25°C, illustrating a further point with regard to development of predictive models, namely that environmental parameters are often correlated with each other, and thus an increase in temperature may enhance the pathogenesis of a pathogen even under conditions of low RH.

Leaf Wetness and Temperature

Research conducted by Aegerter et al. (2003) clearly demonstrates the link between leaf wetness (caused by elevated RH) and temperature with regard to disease incidence. Using *Peronospora sparsa*, the causative agent of rose downy mildew, the experimental results confirmed the commonly held principle of the effect that this association between RH and temperature has on disease incidence and severity. Their data showed a significant effect ($P < 0.001$) of these two environmental parameters on disease severity. Disease severity increased when both temperature and leaf wetness increased, with the highest disease severity observed at 25°C with just 4 hours of leaf wetness. In addition, these researchers were able to demonstrate the effect of RH and temperature on the latent period of infection. The latent period was significantly affected ($P < 0.001$) by temperature, with no immediate symptoms of sporulation of *P. sparsa* occurring at 5°C, but when the samples were kept at temperatures of 22°C, sporulation was detectable after 10 days. Data from these trials have been used to describe the effect of temperature on latent period in some detail (Table 5.3).

Table 5.3 Effect of temperature on latent period.

Inoculation Temperature (°C)	Latent Period at 22°C (days)
5	10
10	8
15	5
20	4
25	4

From these data, Aegerter et al. (2003) were able to develop three logistic regression models that could eventually be developed into a disease-forecasting system for rose growers in the San Joaquin Valley, California. These models are far from complete, but the most promising one incorporates three environmental variables as predictors: (1) the 10-day cumulative number of hours of leaf wetness when temperatures were above 20°C; (2) the 10-day cumulative number of hours when temperatures were in the range of 15–20°C; and (3) the 10-day cumulative number of hours when temperatures were above 30°C. Using the above model, growers would be able to predict the probability of new infections appearing on a given day. However, caution is needed, as during evaluation of the model it was found that it would give false-positive predictions (that is, stating that a new outbreak of disease would be evident on a given day when in fact no new disease was observed) even though the model was statistically a good fit (tested using the Hosmer–Lemeshow goodness-of-fit test), having a chi-square value of $P = 0.99$ (2 degrees of freedom), indicating a good fit of all the data to the model. All of the parameters in the model were statistically significant ($P = 0.05$). Living systems do not always comply with our expectations!

Light Intensity

The response of pathogens to light is another important biological component of epidemiology. Although it is perhaps not as fundamental as temperature and RH in describing disease epidemics, it is nevertheless an integral element of microorganism biology. Pathogens, like most other organisms, use light to detect their environment. During seasonal and daily changes, light functions as an environmental trigger that induces physiological changes. In deciduous plants such as temperate trees, reduced day length in combination with declining seasonal temperature induces changes in leaf biochemistry and consequently visible changes in leaf color (the onset of autumnal leaf color, such as the yellow, russet, and red colors as opposed to the vernal greens of spring and summer). Pathogens also use light as an environmental trigger to mediate several important life-cycle stages, including virulence and pathogenesis.

In *Botrytis cinerea*, the causative agent of numerous gray mold diseases in many crops, changes in the light spectrum have been shown to induce life-cycle changes. For example, in the light, *B. cinerea* produces conidiophores and conidia which are major sources of airborne inoculum. In the dark, *B. cinerea* produces sclerotia, which are a resting stage in the pathogen's life cycle that is resistant to environmental degradation. The perception of light by pathogens, including *B. cinerea*, has also been demonstrated to be a major component of virulence. In *B. cinerea* there is clear evidence that a light perception complex known as the **white collar complex** (WCC) is an essential element in pathogenesis. In this species, pathogenesis and virulence have been found to be correlated with diurnal fluctuations in light and the presence of the WCC in the wild strain (denoted as B05.10) and the absence of the WCC in a WCC-deficient mutant strain (denoted as $\delta\,bcwcl1$). Alternating periods of light and dark induced pathogenesis in the wild type, whereas in continuous dark periods there were no significant differences in lesion size between the wild type and the WCC-deficient mutant (Figure 5.3).

Similar responses to light have also been observed in many other plant pathogens. In the case of the rice blast pathogen (*Magnoporthe oryzae*, also known as *M. grisea*), constant light suppresses disease development, and again this is mediated by the WCC. *Cercospora zeae-maydis*, the causative agent of gray leaf spot disease in maize, infects host plants through the stomata and requires WCC for both appressorium and lesion formation. Clearly, the perception of light is an important biological component of any epidemic, because light controls changes in both the life cycle of a pathogen and its capacity for pathogenesis—two fundamental elements of disease epidemics.

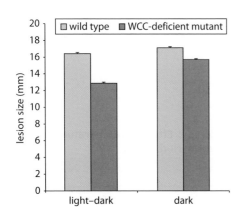

Figure 5.3 Effect of light on lesion size for *Botrytis cinerea*. Note the statistically significant difference between the wild type and the WCC-deficient mutant for the light–dark regime.

Temporal and Cultural Impacts

The manipulation of the cropping environment by growers has a huge impact on plant pathogens, and simple changes in crop rotations have been known to reduce the incidence of disease. For example, the development of take-all (*Gaeumannomyces graminis* var. *tritici*) in wheat can be checked if wheat is followed by a different crop type, such as barley or oilseed rape (OSR). Take-all is a soilborne pathogen that infects the roots and stems of wheat plants, causing a decline in plant health and eventually the presence of white heads in the crop during ear maturation (Figure 5.4). However, take-all is only problematic after a first, or in some cases a second, wheat crop in a rotation. Following a second wheat crop, take-all can give rise to debilitating epidemics in wheat with the potential to reduce yields by up to 50%. However, if a grower has the correct soil type and business model, continuous cropping of wheat is an attractive strategy, so perseverance with continuous wheat has given rise to the phenomena of take-all decline. After 5–6 years of continuous wheat, it has been noted that take-all goes into decline and further cropping of wheat in the same site is possible without the negative impacts of take-all on crop yield. Following take-all decline, wheat yields return to around 90% of those achieved in first or second wheat.

Rotations are an important tool in the control of plant disease, and will be discussed in greater detail in Chapter 10.

Seasonality has a major influence on disease epidemics, and is strongly correlated with climatic variables. The UK is known for its mild and wet climate, and in the summer of 2012 it experienced the wettest summer on record, and consequently yields of all crops were down. In most cases this was due to persistent epidemics of foliar diseases. The UK wheat yield was down by 25%, and much of this was due to continuous reinfection of foliage by pathogens such as *Septoria tritici* and the ever-present mildews. Furthermore, the winter of 2013 was very wet and mild, and the wheat-growing season of 2014 was therefore plagued by high levels of *S. tritici*, with high levels of disease incidence observed on wheat plants early in the spring. The consequence of the wet, mild winter of 2013 was that high levels of inoculum survived the winter as either mycelia or pycnidia on plant debris. In addition, the winter temperatures in 2013 were very mild and seldom dropped below 10°C, the minimum temperature range at which the pathogen is active. Once the pycnidia become wet they produce conidia in long tendrils that can be rain splashed onto young winter-sown wheat plants.

The effect of an unchecked early-season outbreak of disease on crops can be devastating, and may lead to complete collapse of the crop. For example, an unchecked early-season outbreak of mildews on wheat or barley will result in reduced **tiller** survival, and grain quality is compromised as kernel weight is reduced. Indeed, the final yield loss from an unchecked early-season mildew infection can be as much as 35%. However, if a disease occurs late in the season, in particular as the crop is approaching maturity and harvest, the effect of the disease on crop yield is significantly less than that of an early-season outbreak. In dry years, mildew may not establish on a crop at all, or may only partially affect the foliage late in the life cycle of the crop, and hence the impact on yield is reduced because the plant has escaped the ravages of early-season disease and has been able to complete its life cycle unchecked by foliar pathogens.

Figure 5.4 Example of a white head in wheat due to take-all disease.

5.2 Host Factors

Plant Reproduction and Cultivation Material

The type of reproduction and plant propagation system has an impact on the spread and virulence of a disease, both in space and over time. For example,

Table 5.4 **Examples of key diseases of potato tubers.**

Common Name	Scientific Name	Parts of Plant Infected
Late blight	*Phytophthora infestans*	Foliage (known as the haulm) and tubers
Potato common scab	*Streptomyces scabies*	As spots and corky tissue on the surface of tubers; may also appear on roots and stolons
Potato powdery scab	*Spongospora subterranea* f. sp. *subterranea*	Tubers, roots, and stolons
Potato dry rot	*Fusarium* species (*F. solani* var. *coeruleum* and *F. avenaceum* are common)	Tubers, and persistent in the soil for up to 9 years
Potato watery wound rot	*Pythium ultimum*	Tubers, and in the soil as an oospore, which can then re-infect tubers at harvesting
Potato gangrene	*Phoma* species	Haulm and tubers
Potato pink rot	*Phytophthora erythroseptica*	Haulm and tubers
Potato black leg	*Erwinia carotovora*	Stems and tubers
Potato silver scurf	*Helminthosporium solani*	Tubers
Potato skin spot	*Polyscytalum pustulans*	Shooting sprouts and tubers

potatoes (*Solanum tuberosum*) are grown from vegetative seed tubers which have to be certified disease-free. Tempting as it might be for growers to save on the cost of buying seed potatoes, this is an unwise strategy because saved seed potatoes, if not certified disease-free, are quite likely to have a number of overwintering sources of inoculum from a range of common and debilitating pathogens (Table 5.4). These pathogens will infect the next season's potato crop if that crop is grown from non-certified seed potatoes. Vegetative material used for propagation is always problematic, and the grower must ensure that this material is disease-free or they run the risk of building up levels of disease inoculum on their estate, which will cause persistent problems in successive rotations. This is pertinent to crops grown from tubers or bulbs, as the harvest rate is never 100% and therefore volunteer tubers or bulbs will be present, which if infected will act as a source of inoculum. Even if the next potato rotation is 2–3 years away (for example, potato → wheat → oilseed rape → potato), the persistent volunteers will act as a focus point for the potato crop in the fourth rotation in this example.

Another good example of problems caused by vegetative propagation stock comes from the ornamental sector. Spring bulb production in Europe is an economically important trade, especially with regard to stalwarts such as the daffodil crop (*Narcissus* species) and the tulip crop (*Tulipa* species). Both of these examples rely on the production of flower stems and bulbs from vegetative propagation material that can take 2–3 years to reach flowering maturity. This is ample time for the crop to become infected with numerous diseases that will persist on the bulbs from year to year. A serious disease in daffodils is basal stem rot (*Fusarium oxysporum* f. sp. *narcissi*), which overwinters on stored bulbs and then in growing plants the following season. The pathogen causes premature **chlorosis** of the leaves, and can lead to blind (non-flowering) bulbs. In bulbs that are stored, rotting will eventually occur. Clearly if a grower sells to the garden and amenity sector infected bulbs that have persistent sources of disease inoculum, that grower will lose all their business credibility. Therefore the crop must be certified disease free, just like the seed potatoes.

Fruit growers also have persistent problems with sources of overwintering inoculum. This includes perennial cane fruits such as raspberries and blackcurrants, as well as top fruits such as apples and pears. In the case of cane fruits, stems infected with *Botrytis cinerea* can become a focal point for the infection of new shoots in the next growing season. For apples, one major source of inoculum is fallen apples left on the floor of the orchard. For example, brown rot (*Monilinia fructicola*) can overwinter as sclerotia in mummifed fruit (Figure 3.8). At the onset of spring the sclerotia in the mummified apple will produce new conidia that can be transmitted via wind and rain splash back onto the new fruits.

Plants that are grown from seed tend not to suffer from persistent infections from generation to generation like those observed in vegetative propagation material. This is because many fungal infections are isolated from the seed during the seed maturation process. However, there are a few exceptions, such as the loose smut (*Ustilago nuda*) of wheat and barley, and common bunt of wheat (*Tilletia caries*). Both of these pathogens persist in the seed as hyphae and at the teliospore stage. At harvest, the teliospores are released to the environment, and can then be blown significant distances due to their position high up in the canopy. Alternatively, the germinating teliospores can infect the maturing embryo, resulting in hyphal penetration of the seed. Again, certified seed has helped to control the level of these two pathogens in our modern cropping systems. This highlights an important concept in relation to modeling plant disease, namely the effect of sanitation on the progress of a disease. This concept will be illustrated in the section on modeling of plant disease epidemics.

Race-Specific Issues

The gene-for-gene theory was briefly introduced in Chapter 4, but here we shall explore this concept further in relation to race-specific issues associated with epidemiology and subsequently pathogen control strategies. In race-specific issues the concept of the gene-for-gene theory can be applied to the modeling of disease epidemics. In 1911, Redcliffe N Salaman, a pioneering potato breeder, conducted some of the early research in this area when he looked at naturally occurring resistance of the wild potato plant (*Solanum demissum*) to the late blight pathogen (*Phytophthora infestans*) (Research Box 5.1). This early work concluded that *S. demissum* escaped late blight because it contained genes that conferred resistance to late blight. Salaman later went on to write a number of associated publications on potato breeding and genetics. The genes that conferred resistance were termed R_1, R_2, and R_3, and potatoes that contained one or more of these genes were resistant to the common race of late blight termed r_0. The races of the late blight pathogen that could infect potato cultivars with the associated R genes (R_1, R_2, and R_3) were termed r_1, r_2, and r_3, respectively. Continuing the story of the potato crop and this associated nemesis, the late blight pathogen, brings us to the development in the UK of a main crop potato known as Pentland Dell. This variety further illustrates the effect of race-specific issues. Pentland Dell contains a suite of resistance genes identified as R_1, R_2, and R_3, and at the time of its release to growers in the early 1960s, no race of *P. infestans* was compatible with the crop. However, by 1967, Pentland Dell was being widely grown in the UK, and resistance to *P. infestans* broke down with the emergence of races of *P. infestans* termed r_1, r_2, and r_3. The development of disease-resistant varieties of crop plants is a major strategy used by growers across the globe. Such **vertical resistance**, as it has been termed by Van der Plank (1963), still forms the basis of plant breeding programs for all our major food crops today.

Plant Age and Susceptibility to Disease

As plants grow, develop, and eventually age, their susceptibility to pathogens changes. In the very early stages of life, particularly as emerging seedlings, plants are exceptionally vulnerable to a range of damping-off diseases,

RESEARCH BOX 5.1 EPIDEMIOLOGY OF *PHYTOPHTHORA* DISEASES IN TREES

Phytophthora diseases have been known to growers for over 150 years, and *Phytophthora infestans* has been linked to mass migrations of people from Ireland, Wales, and the Highlands and Islands of Scotland in the 1840s. *P. infestans*, the causative organism of potato blight in the 1840s, caused widespread potato crop losses on an epidemic scale in the western regions of the UK, and hence was one of the major contributors to rural poverty and subsequent migrations of over a million people.

There is now a new global threat to plant health, and specifically to the health of our trees, from these oomycota pathogens of the *Phytophthora* genera. Currently 70 species of tree pathogens belonging to the genus *Phytophthora* have been described, and it is predicted that there are numerous undescribed species from this genus in isolated natural forest ecosystems around the globe. Unlike many of the classically described epidemics of the crop pathogens, the epidemiology of this tree disease is somewhat different, and in many cases driven by anthropogenic processes such as exploitation of plants and the global trade in trees and wood products. Trading and movement of trees across the globe often occur through the nursery sector and the subsequent landscape planting of trees. All of this activity has been undertaken with minimal biosecurity controls (see Chapter 10), which has given rise to major outbreaks of *Phytophthora* diseases in the USA, Europe, Australia, and New Zealand. Furthermore, because of the mixing of tree stock within nurseries, reports of new and evolving *Phytophthora* pathogens continue to cause significant problems with regard to the health of our global forests and tree stocks.

Epidemiology of Contemporary *Phytophthora* Pathogens

There is a very long list of *Phytophthora* pathogens that infect many different tree species across the globe. However, there are a few species that are currently causing major problems in our forests, and many of these pathogens infect numerous tree species (Table 1). The epidemiology of these *Phytophthora* pathogens is complex, as they have complex life cycles (see Chapter 9), giving rise to several stages of potential inoculum. Consideration of the ways in which humans have carelessly exploited plants in recent years shows that this has given rise to a number of factors associated with the epidemiology of these diseases. These include the following:

- Soil- and root-based inoculum
- Soil and root to aerial transmission
- Aerial inoculum and secondary host species
- Nursery-based inoculum
- Transmission via flood water
- A complex array of infectious *Phytophthora* species
- Evolution of new *Phytophthora* species
- Transmission by foot fall and dog walkers
- Climate change and shifting patterns of *Phytophthora* species.

It is the aim of this research box to explore a range of the above-mentioned issues in relation to the epidemiology of the genus *Phytophthora*, and to reflect on how an understanding of these highly infectious diseases may aid their control.

Soil–Root and Soil–Root to Aerial Transmission of Inoculum

Some *Phytophthora* species, such as *P. cambivora*, *P. citricola*, and *P. gonapodyides* (Table 1), are generally considered to be pathogenic as **non-caducous spores**, which are typically root- and soil-based spores. Trees infected with this type of spore will exhibit symptoms of collar and root disease, which lead to a general decline in the health of infected host trees. In 2005, aerial lesions were reported on host trees such as beech (*Fagus sylvatica*) infected by the non-caducous soil and root spores of *P. cambivora* and *P. citricola*. Aerial lesions in the form of bleeding bark lesions have also been noted on European sycamore (*Acer pseudoplatanus*) and horse chestnut (*Aesculus hippocastanum*) infected by non-caducous soil- and root-based spore stages of numerous *Phytophthora* species. The means by which these generally soil-based spores are translocated to the canopy is difficult to explain, but it seems most likely that the non-caducous soil-based spores are transmitted to the canopy by foraging snails. Researchers have isolated oospores from fresh exudates of beech infected with *P. citricola*, and it has been subsequently demonstrated that snails will feed on these exudates (Jung, 2005). Another theory is that soil- and root-based spores of *P. cambivora* and *P. citricola* are transmitted to the canopy via embolisms in infected xylem tissue. A third possibility is that spores of *P. gonapodyides* may be present in the pools of water that collect at the junction between major branches and the trunk of beech. It is clear that regardless of the mechanisms of transmission of soil- and root-based spores to the canopy, the result is that these diseases have an additional aerial phase, which will of course enhance their spread through a forested landscape.

Sources of Inoculum

Most *Phytophthora* pathogens of trees are typically soil-borne, but in 2003 a new *Phytophthora* species, *P. kernoviae*, was detected in *Rhododendron* and beech (*Fagus sylvatica*) in Cornwall, England. The disease was initially found in woodlands and gardens, and in a small number of nurseries in the county. Subsequently, in 2007, *P. kernoviae* was isolated on bilberry (*Vaccinium myrtillus*) in woodlands across Cornwall, and in 2008 on bilberry on open heaths across the county. This pathogen, along with another devastating tree pathogen, *P. ramorum*, is spread locally via rain-splashed spores from infected needles and leaves to uninfected material. However, for these two diseases a more damaging mode of dispersal is the wider movement of spores via wind-borne mists, which has the potential to increase the distribution of both pathogens across a region relatively quickly. This has been observed for *P. ramorum* in the canopies of oak, and was regarded as the driving force for outbreaks of sudden oak death in the Americas. Another mode of dispersal for these two pathogens, with the potential to cause serious damage, is via watercourses and irrigation water. In addition, both pathogens have a resting chlamydospore stage, during which the spores can survive for a number of years (at least 3 consecutive years in contained

RESEARCH BOX 5.1 EPIDEMIOLOGY OF *PHYTOPHTHORA* DISEASES IN TREES

Table 1 Summary of tree diseases associated with *Phytophthora* species, including hosts, secondary hosts, ecological notes, and distribution.

Species of *Phytophthora* Pathogen	Main Host Trees and Shrubs	Main Secondary Hosts	Ecological Notes	Distribution
Phytophthora taxon *Agathis*	*Agathis australis*	Unknown	First reported in 2006. Motile spores swim through wet soils and infect the roots of lauri trees	New Zealand
Phytophthora alni subsp, *alni*, *uniformis*, and *multiformis*	*Alder glutinosa*, *A. incana*, and *A. cordata*	In the wild only the *Alnus* species have been reported to be susceptible, but the results of greenhouse experiments indicate that *Juglans regia*, *Castanea sativa*, and *Prunus avium* may also be susceptible	Thought to be a hybrid between *P. cambivora* and *P. fragariae*, and now spreading as a hybrid swarm across Europe. Spread and infection occur mainly via soil- and water-borne zoospores; thus widespread transmission is possible along riparian strips and planting	Austria, Belgium, France, Germany, Hungary, Ireland, Italy, Lithuania, Netherlands, Sweden, and UK
cactorum	*Aesculus hippocastanum* and *Fagus sylvatica*	A further 200 species of trees and ornamental shrubs, fruit crops	Complex multi-stage life cycle. Able to survive as hyphae and as chlamydospores and oospores. Zoospores swim to a host and infect the root tips or the vascular system of the crown	Austria, Germany, Italy, Sweden, and UK
cambivora	*Acer pseudoplatanus*, *Castanea sativa*, and *Fagus sylvatica*	*Rhododendron*, *Pieris* species, *Abies procera*	Non-caducous soil- and root-borne spores. There are new reports of non-caducous spores as aerial inoculum	Austria, Germany, Italy, Sweden, UK, and Oregon and North Carolina
citricola	*Aesculus hippocastanum*, *Fagus sylvatica*, and *Juglans nigra*	This pathogen has a wide host range, but is a major threat to avocado (*Persea americana*) crops. A review by Oudemans et al. (1988) describes an increasing threat to avocado growers in California	As above, but the main phases of the life cycle are soil borne. Initially the pathogen attacks the roots, so there is not above-ground expression of the disease. As the disease progresses into the feeder roots, sporangia are produced which release zoospores that are able to freely swim in wet soils to neighboring trees	Wide distribution including Europe, UK, USA, Asia, and Australia
gonapodyides	*Alnus glutinosa*, *Fagus sylvatica*, and *Quercus ilex*	*Malus* species (apple trees), seedlings of the family Pinaceae, including *Pseudotsuga menziesii*, *Abies* species, and *Tsuga mertensiana*	Non-caducous soil- and root-borne spores. There are new reports of non-caducous spores as aerial inoculum. Tolerant of high temperatures and favors riparian zones	Austria, Belgium, Germany, Hungary, Italy, New Zealand, Patagonia, Poland, Spain, and USA
kernoviae	*Aesculus hippocastanum*, *Fagus sylvatica*, and *Quercus* species	*Rhododendron*, *Ilex* species, *Vaccinium myrtillus*, and numerous other hosts	Leaf-to-leaf and plant-to-plant infection from rain-splashed zoospores. Oospores can reside in the soil and be transmitted by animals and footwear	South-West England, particularly Cornwall
lateralis	*Chamaecyparis lawsoniana* and *Taxus brevifolia*	*C. formosensis*, *C. obtusa*, and possibly *Thuja occidentalis*	Zoospores swim through wet soil and surface water and infect root tips. Vertical spread up the trunk is limited to about twice the diameter of the stem. Zoospores also move through watercourses, facilitating further spread. Aerial infections have been very limited, but were observed in the outbreaks in Brittany which have been attributed to deciduous sporangia	Oregon, Northern California, Canada, France, and the Netherlands; Glasgow, Scotland in 2010, and County Down, Northern Ireland in 2011

(Continued)

RESEARCH BOX 5.1 EPIDEMIOLOGY OF *PHYTOPHTHORA* DISEASES IN TREES

Table 1 (*Continued*) Summary of tree diseases associated with *Phytophthora* species, including hosts, secondary hosts, ecological notes, and distribution.

Species of *Phytophthora* Pathogen	Main Host Trees and Shrubs	Main Secondary Hosts	Ecological Notes	Distribution
pseudosyringae	*Alnus glutinosa*, *Fagus sylvatica*, and stems of *Lithocarpus* and *Quercus* species	Fruit rot in *Malus* species (apple trees) and leaf rot in *Ilex aquifolium* when inoculated in the laboratory. As *I. aquifolium* is strongly associated with *Quercus* species, observations of *P. pseudosyringae* in *Ilex* in the wild may become more common	Soilborne with motile spores	Austria, Germany, Italy, Sweden, UK, California, and Oregon
ramorum	Numerous species of *Quercus*, *Fagus sylvatica*, *Rhododendron ponticum*, *Nothofagus obliqua*, *Acer pseudoplatanus*, and *Aesculus hippocastanum*	*Rhododendron*, *Viburnum*	Zoospores on leaf material are locally spread by rain splash and wind. Longer-distance transmission is most probably through movement of contaminated material, either in the growing media or on vehicles, footwear, or animals	Canada, Europe, and west-coast USA. Has been introduced to east-coast USA via nursery stock
syringae	*Fagus sylvatica*, *Syringa vulgaris*	Numerous hosts from 29 genera across 14 families. Main crop infected is *Malus* species (apple trees), in which *P. syringae* causes fruit rot	Overwinter as resting oospores, often from fallen or discarded apples. In the spring and summer the resting spores germinate (optimum temperature range is 10–14°C), producing motile zoospores that can be splashed onto foliage	Africa, Argentina, Asia, Australasia, Brazil, Canada, Europe, and USA

quarantine experiments) on plant debris and inoculated soils. The subsequent movement of plant debris and infected soil from an infected site to an uninfected site via vehicles, footwear, plant debris (infected needles and leaves), and nursery stock has been implicated as one of the most widespread causes of dispersal of these two *Phytophthora* pathogens in the UK, and of *P. ramorum* from the west coast to the east coast of North America.

Nursery Stock

The widespread movement of plant material across borders and within a country has had profound implications for the dispersal of *Phytophthora* pathogens. Early evidence of the effects of this damaging practice was found in both oak and alder trees. In 2003, *P. ramorum* in oak was observed to be spreading from nursery stock to native *Quercus* species in Cornwall. In addition, there is evidence that the bleeding stems of imported American oak species have infected native oaks and beech in the UK. With regard to the alders, studies have shown that the disease caused by *P. alni* is common along riparian strips in the southern UK, and in France, Germany, and a number of other European countries. Further evidence of the involvement of nursery stock comes from

an outbreak of *P. alni* in Bavaria. There the majority of the observed outbreaks were in non-flooded forest stands that had been planted up from nursery-grown stock (92% of the stands), and extensive field investigations found that 58 out of 60 river systems investigated showed that the source of the inoculum could be traced back to young, infested, riparian alder strips and young alder plantations, where stock originated from nurseries. The role of nurseries in the transmission of *P. alni* inoculum was further corroborated by baiting tests on rootstocks of alder across the nursery sector, and demonstrated that rootstocks from three out of four nurseries were infected with *P. alni* where the nurseries brought in clonal rootstocks for resale. No inoculum was isolated from nurseries that grew trees from seed.

A further issue with regard to *Phytophthora* and the nursery sector is that the recombination of numerous *Phytophthora* species within nurseries, particularly large woody plant nurseries where susceptible species such as the rhododendrons are grown, has led to the evolution of new species of *Phytophthora*. Molecular analysis has shown that the evolution of *P. alni* is the result of a cross between *P. cambivora* and *P. fragariae*, and so it is basically a hybrid. The number of other new *Phytophthora* hybrids that are evolving is unknown and is a cause for concern in the forestry and horticultural sectors, demonstrating the need to improve biosecurity controls across both sectors.

A Complex Array of *Phytophthora* Species

A further complication when modeling the epidemiology of *Phytophthora* pathogens associated with tree diseases is the observation that in some cases diseased trees may have a complex array of *Phytophthora* species associated with a disease outbreak. Another cause for concern is the evolution of the concept of the hybrid swarm of *Phytophthora* species, as observed in the pathology of infected *Alnus* species. Sweet chestnut (*Castanea sativa*) is susceptible to *P. cinnamomi* and *P. cambivora*, giving rise to a disease known as ink disease, or collar and root disease. These two pathogens have caused significant epidemics in southern Europe in both the nineteenth and twentieth centuries. The disease remains serious today, and as a result a major EU-funded project (CASCADE II, 1999 to 2006) surveyed 35 sites across southern Europe in order to characterize the number of *Phytophthora* species associated with stand dieback of sweet chestnut. The results of this survey indicated that in soil samples from declining stands of sweet chestnut a further five species of *Phytophthora* pathogens could be isolated and characterized, namely *P. cactorum*, *P. citricola*, *P. megasperma*, *P. cryptogea*, and *P. syringae* (Vettraino et al., 2001).

Summary: Reflections on *Phytophthora* Infections in Tree Species

It is clear, from the wealth of data in the literature on the complex nature of *Phytophthora* diseases attacking the tree flora, that the development of a single disease prediction model is a complex process and that it will take some years to develop. Once a model has been developed, further research will be required to evaluate its accuracy using both computer simulations and long-term monitoring, because of the numerous variables involved in modeling the *Phytophthora* diseases of trees. Due to the complex nature of these *Phytophthora* pathogens and the array of host and secondary host species (Table 1), continued detailed research and monitoring will be required in order to compile an exhaustive list. If modelers were to consider only one of the *Phytophthora* species discussed above, say *P. ramorum*, although there are considerable data on the suite of species susceptible to this pathogen, it is also quite likely that this dataset is incomplete, which would compromise the development of disease forecast and progression models.

In Europe there is an ongoing project to develop a disease forecast model for *P. ramorum*, known as the Risk Analysis for *Phytophthora ramorum* (RAPA) project (Sansford et al., 2010). This has highlighted a number of key steps towards the devel-opment of a forecast model, and has been able to evaluate the risk posed by *P. ramorum* with regard to entry into Europe from exotic lineages and mating types of the pathogen. It has also been able to develop a distribution map for *P. ramorum*, based on research initially conducted in the USA. The distribution map is for natural and semi-natural environments, and is based on a ranking of climatic factors that favor the spread of the pathogen, in this case mild wet winters. However, the ability of the pathogen to spread from the long-lived chlamydospore stage also makes it suitable for persistence in the Mediterranean region. The model predicts that central, western, and southern Europe will be most vulnerable to the spread of the pathogen. However, these predictions will not be consistent under conditions of climatic warming. Another weakness of the proposed model is the unknown quantity posed by the plant nursery sector, particularly with regard to cultural practices within plant nurseries, such as above-ground irrigation (mist-blown spores), use of contaminated irrigation water (there are numerous examples of *P. ramorum* and other *Phytophthora* pathogens surviving in contaminated irrigation water), and the use of fungicide compounds across the nursery sector (particularly the use of metalaxyl-M, as there have been reports of fungicide resistance building up in the *P. ramorum* population). Another factor linked to the impact of nurseries on the above-mentioned model is the importation of new stock into Europe, which may result in new host species becoming available for the pathogen.

Finally, a weakness of any disease prediction model for natural and semi-natural environments is the fact that these environments have numerous access points. For instance, in the UK the "Right to Roam" act (2005) and the existence of many other open-access areas mean that people can freely access a range of sites, such as moorlands, heaths, coastal paths, and woodlands. These sites contain a number of key plant communities that are composed of plants susceptible to *P. ramorum* and its close relative *P. kernoviae*, such as *Vaccinium myrtillus*, and currently there are no comprehensive biosecurity protocols for walkers to follow. Indeed, one internationally important woodland site within the East Dartmoor National Nature Reserve has open access and is a favorite site for dog walkers and educational bodies from across Devon and Cornwall, but there is no public information about the threat of *Phytophthora* pathogens, or any form of biosecurity control, at the many points of access. Thus any meaningful predictions about the development and spread of *P. ramorum* derived from the above-mentioned model would be significantly compromised by uncontrolled public access—exactly the same scenario as caused the initial outbreak of these very damaging diseases in the first place.

such as *Rhizoctonia solani*, which can cause root rots, and the *Pythium* species, which are very damaging oomycetes. However, as plants establish and mature they become less susceptible to damping off. For example, very young seedlings of bean (*Vicia faba*) are highly susceptible to *Pythium ultimum*, which can cause complete collapse of an emerging crop. Conversely, after 6 days of post-emergence growth, *P. ultimum* may still cause root necrosis

Figure 5.5 An ancient oak with a range of crown galls.

but does not result in collapse of the crop. The same observations can also be made for the biotrophic mildews, where young plants are far more vulnerable to these debilitating foliar pathogens, whereas mature plants are better able to cope with the mildews, and still function and produce some yield. As plants age they may acquire high levels of pathogens, and trees in particular are known to harbor high levels of inoculum of many pathogens. The most obvious of these is the crown gall disease (*Agrobacterium tumefaciens*) seen in many deciduous trees, such as oak (*Quercus* species). This pathogen causes spectacular deformations in growth form (Figure 5.5) and, as a result, aged trees that have been infected for many decades appear like characters from many a fantasy novel!

Accumulation of high levels of pathogenic inoculum can be very damaging to cropping systems that are based on clonal propagules, such as the banana crop (*Musa* species). In Kenya, high levels of the infectious and debilitating virus known as banana streak virus (BSV) caused serious problems for subsistence farmers who relied on farm-saved vegetative clones. This problem was only alleviated following the eradication of the virus in vegetative clones by the application of tissue culture and subsequent stock certification and farm extension programs. Similar issues face potato growers in the UK, where many seed tuber pathogens, often originating from the vegetative culm, such as late blight (*Phytophthora infestans*), can be transmitted to the new crop. Again the standard practice for potato growers is to use certified disease-free seed tubers.

Another factor relating to plant age and pathogenesis, and one that is often overlooked, is the concept of **ontogenic resistance**. This is a form of disease resistance that increases with age, not only at the whole-plant level but also in components of plants, such as leaves and fruits. One excellent example from nature is seen in the seedlings of dipterocarp seeds (Figure 5.6). Dipterocarps are one of the main tree families of the tropical forests of South-East Asia, and have very complex reproductive systems. Basically dipterocarps synchronize their flowering and are known as **masting** trees, producing many thousands of seeds when they eventually flower. Once these seeds reach the forest floor, they germinate quickly, typically 3–5 days after being shed from the mother tree, and in the very early stages of their growth they are extremely vulnerable to predation and disease. However, as they develop into young seedlings they quickly produce a range of **secondary metabolites** that protect them against both predation (herbivory) and subsequent infection by numerous foliar pathogens. This is an essential physiological development for the dipterocarp seedlings, as they spend 15–20

Figure 5.6 Dipterocarp seedlings in a quiescent stage on the forest floor at Danum Valley in Sabah, Borneo.

years or more at the seedling stage (about 1.2 m in height). This is because the forest canopy is obviously dominated by their parents, and the seedlings are waiting for a gap in the canopy before they switch from their quiescent seedling life form to an actively growing tree, racing for the canopy.

Returning to cropping systems, ontogenic resistance has been observed in grape fruits, where the young fruits are very susceptible to grape powdery mildew (*Uncinula necator*), but after 3–4 weeks the grape berries develop age-related resistance to powdery mildew. Grapes may be infected, but germinating conidia of the powdery mildew fail to establish haustoria and sub-cuticle hyphae. The detailed mechanism of this age-related resistance is not yet fully understood. Further evidence of age-related resistance in 3- to 4-week-old grape fruits is an 18% increase in the latent period of the mildew infection compared with the younger fruit. However, a note of caution is needed. Although 4-week-old grape fruits may be resistant to *Uncinula necator*, the presence of propagules on the fruit will lead to secondary and post-harvest infections if adequate control strategies are not employed. Thus, in relation to epidemiology, a fundamental requirement when predicating an epidemic and designing a disease control strategy is the alignment of control measures with the actual risk of infection, which of course could change as plants age. This is particularly pertinent to top fruit crops such as apples (a topic that will be discussed in detail in Chapter 10).

The above-mentioned climatic conditions and anthropogenic variables typify many cropping systems as growers select their crops (both species and variety) to match the markets, and time their cropping to match the prevailing growing season. Furthermore, the close spacing between plants in a crop will favor spread of the pathogen via wind and rain splash of conidia. Modeling of these dynamic parameters is under constant academic review, and new models are being developed all the time. In the context of this chapter a review of two basic models is warranted, namely the simple-interest and compound-interest models, both of which will give the student the fundamental starting points for exploring this complex and growing area of academic and applied research. However, before we examine the mathematical elements of epidemiology we must first consider some of the pathogen-related factors that are also key components of any given epidemiological model.

Pathogen Factors

Pathogen life cycles are complex, involving a range of biological structures, such as spores and overwintering **sclerotia**, and reproductive strategies. Furthermore, some pathogens may overwinter on **secondary hosts**, such as the berberis shrub which is used by the rust pathogen of wheat. The complex life cycles of pathogens are explored in more detail in each of the compendia (Chapters 6, 7, 8, and 9). However, to aid understanding in relation to plant epidemiology, two life cycles are presented here. A typical *Fusarium* wilt will be used as an example of the life cycle of a **monocyclic** pathogen, and *Monilinia fructicola* (brown rot of apple) will be used to illustrate the life cycle of a **polycyclic** pathogen (Figure 5.7).

Many pathogens have a simple rate of replication in relation to a given crop or cropping season. These are known as monocyclic pathogens, because they only replicate once within a single growing season. Such pathogens tend to be soilborne diseases, and they infect a crop as they come into contact with it. Once the crop debris has decayed, spores of monocyclic pathogens are released into the field ready to infect new hosts. Examples of monocyclic pathogens can be found among the soilborne pathogens of the genera *Fusarium*, *Verticillium*, and *Plasmodiophora*. One can relate this monocyclic life cycle mathematically to the accumulation of simple

Figure 5.7 Schematic diagrams of the disease life cycles of (a) *Fusarium* **wilt, a monocyclic pathogen, and (b)** *Monilinia fructicola* **brown rot of apple, a polycyclic pathogen.** (Adapted from Agrios G [2005] Plant Pathology, 5th ed. Elsevier Academic Press.)

(a)

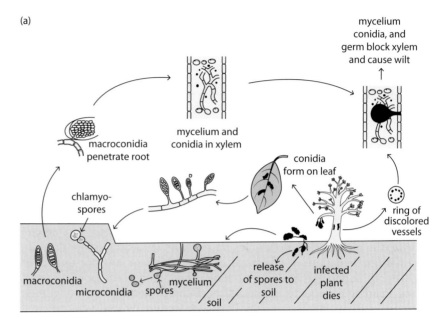

(b)

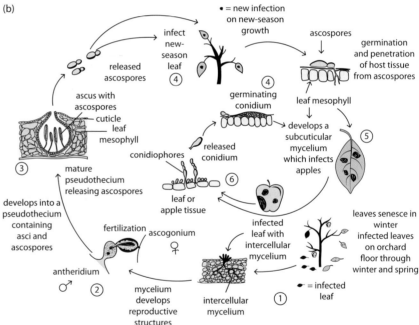

interest in banking terms. This is an ideal concept, as it allows epidemiologists to describe the rate of disease increase in any given season. Caution is needed, however, as we shall see that natural systems are not as straightforward as this.

With regard to modeling disease epidemics, the next major life cycle strategy used by pathogens is described as polycyclic. Here the pathogen can multiply very rapidly because it replicates several times within a single growing season. The types of diseases that are described by polycyclic replication rates tend to be the foliar diseases, such as the rusts and mildews, and *Septoria tritici* in wheat. *S. tritici* is a very common foliar disease of wheat in the West Country in England. One of the major problems with *Septoria* in wheat is that the early stages of the disease affect only the lower leaves of the crop, but as the crop grows and extends the disease travels up the expanding green tissue through successive cycles of replication and spore dispersal, and can eventually infect the all-important flag leaf (Figure 5.8). This highlights one of the major conceptual problems in epidemiology, namely that rapid increases in levels of disease inoculum occur within a single cropping season. This is known mathematically as exponential growth.

5.3 Modeling Disease Epidemics

The Role of Mathematical Models in Plant Pathology

Van der Plank (1960, 1963) developed some of the early mathematical models used to describe the temporal development of disease epidemics that have since formed the basis for contemporary disease modeling. In epidemiology, modeling aims to characterize the spatial and temporal patterns of disease progress, which can then be used as part of a grower's armory against the ever present threat of devastating epidemics. Some of the first disease models were those developed for potato blight. In 1969, Waggoner and Horsfall published their model EPIDEM for predicting the progress of early potato blight caused by *Alternaria solani*, and this model is still used today.

In order to describe the spatial and temporal characteristics of a disease as it moves through a cropping region it is essential to define some key objectives. These objectives will then help the modeler to establish the kind of mathematical representation that is required. Basically there are three categories of models as described by Kranz and Royle (1978): descriptive, predictive, and conceptual (Table 5.5). The model used to illustrate the concept of epidemiology in this chapter falls into the descriptive and predictive category, and follows the simple formula outlined below:

$$x = x_0 e^{rt} \tag{5.1}$$

where x is the amount of disease at any given time, x_0 is the amount of disease at the start of an epidemic, e is a mathematical constant, r is the rate of increase of the disease, and t is time (adapted from Jones, 1987, after Van der Plank 1963).

Figure 5.8 *Septoria tritici* **on wheat.** A clean flag leaf and an infected flag leaf are shown.

Table 5.5 **Summary of classification system used to describe disease models.**

Category of Model	Objective of the Model
Descriptive	Used to generate a hypothesis or describe the results of disease monitoring. However, these models generally do not reveal the mechanisms underlying the disease progress
Predictive	These models are confusingly also referred to as descriptive. However, predictive models describe the occurrence and severity of the disease and subsequent epidemic. Both descriptive and predictive models use mathematical tools such as regression and differential equations
Conceptual	Known as explanatory or analytical tools. They allow identification of the problem by elucidating the cause and effect of an epidemic. Conceptual models can be used to describe the biological and ecological processes that underpin an epidemic, and hence afford a greater level of understanding by critical evaluation of the ecology of the pathogen. Conceptual models are excellent tools for gaining a theoretical understanding of the generic features of an epidemic, and thus are very useful sources of decision-making strategy for policy makers

Before we can develop this equation further it is necessary to explore some of the ecological concepts that underpin the terms in Equation 5.1. From the above, one of the main components is the rate of increase of the disease (r), and this is related to the life cycle of any given pathogen outlined above. In mathematical modeling the monocyclic pathogen is referred to as the simple-interest organism and the polycyclic pathogen is referred to as the compound-interest organism. Plotting the growth of a polycyclic pathogen over a cropping season will lead to a sigmoid growth curve that illustrates a slow start and lag phase as both the crop and the pathogen establish in the field. Then as the season warms up and the growth of both the crop and the pathogen is accelerated the plot becomes logarithmic or exponential. Eventually, as the crop dies and less material is available for infection, there is a flattening off of the growth curve, the asymptote (Figure 5.9). If the growth of a monocyclic pathogen was to be plotted over one growing season the resultant curve would not be sigmoidal. However, if the growth of a monocyclic pathogen was to be plotted over several successive seasons, eventually a sigmoid curve would emerge.

The observation that polycyclic pathogens give rise to an asymptote is important, as it indicates that we need to modify Equation 5.1 to account for the decrease in susceptible plant material available for continuous infection throughout the growing season. Therefore as a polycyclic epidemic progresses throughout a crop or region its rate of increase is controlled by the amount of disease (inoculum) present, x, and the proportion of susceptible tissue left ($1 - x$). A more detailed discussion of Equation 5.1 can be found in Jones (1987). Basically, however, when plotting the progress of a disease the sigmoidal curve is straightened out by applying a logarithmic correction as follows:

$$r = (1/t_2 - t_1) \log_e (X_2/X_1) \tag{5.2}$$

Suppose that the average percentage of yellow rust on wheat at time 1 is assessed and found to be 0.1% ($X_1 = 0.1\%$), and 10 days later the average percentage of yellow rust on wheat is again assessed and determined to be 2% ($X_2 = 2\%$). As it is known that the total time that elapsed between the first disease assessment and the second disease assessment is 10 days, the mean increase in diseased units per day can be calculated.

Thus: $r = (1/10 \log_e 0.02)/0.001$
             $= 0.2996$ disease units per day.

Again this works well until about 5% of the host tissue has been infected; then the correction factor outlined above must be applied.

One principle that should be obvious from Equation 5.1 is that any disease control strategy which can reduce either the rate of increase of the disease

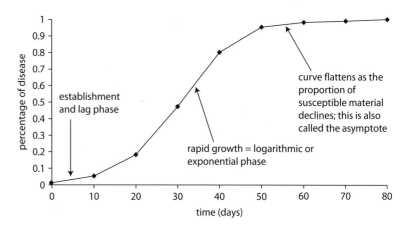

Figure 5.9 An example of a disease progress curve showing the key phases of a typical sigmoidal curve.

(r) or the level of inoculum (X_0) of the pathogen at the start of the disease will help to reduce the final level of disease in a crop. This is known as sanitation, and it can be achieved through a number of measures, including cultural, biological, and chemical control, and plant disease resistance. In the context of this chapter the following equation (5.3) can be applied:

$$\delta t = (1/r)\log_e (X_o \div X_{os})\tag{5.3}$$

where X_o is the level of pathogen (inoculum) with no sanitation, and X_{os} is the level of pathogen (inoculum) after sanitation.

The final outcome of Equation 5.3 is that application of successful sanitation will bring about a delay in the progress of the disease, known as delay time. Substituting some model data into Equation 5.3 illustrates the concept of delay time, which remains constant throughout the exponential phase of the disease (Figure 5.9).

Furthermore, by illustrating the application of Equation 5.3 we can substitute some example data into a sanitation scenario as follows.

First we need to establish the initial level of sanitation. In our example (Figure 5.10 and Table 5.6) the initial level of inoculum was estimated to be 1%, and following application of a sanitation measure the inoculum level was estimated to be 0.1%. This equates to a 90% control of the initial inoculum, leaving a final estimate that there is only 10% of the initial amount of disease after sanitation.

Thus (noting that $\log_e$ is transformed to $\log_{10}$):

$$\delta t = 0.01/0.001$$
$$= 100\log_{10}$$
$$= 2.$$

Multiplying this by 10% gives a delay time of 20 days (Figure 5.10).

The contemporary development of plant disease epidemiology continues to try to elucidate the spatial and temporal dynamics of epidemics, and aims to draw upon the results of these endeavors to inform growers of potential

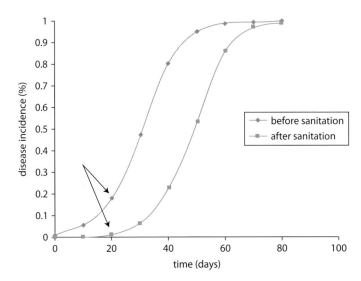

Figure 5.10 An example of disease progress curves showing the effect of sanitation on the delay time. Note the significant differences at a delay time of 20 days, indicated by the two arrows.

Table 5.6 Data for sanitation model.

Days	Initial Level of Inoculum	Level of Inoculum Following Sanitation
0	0.01	0.001
10	0.05	0.001
20	0.18	0.020
30	0.47	0.060
40	0.80	0.230
50	0.95	0.530
60	0.98	0.860
70	0.99	0.970
80	1.00	0.990

Figure 5.11 A schematic example of the spatial patterns of soil- and airborne disease. (a) Soilborne monocyclic pathogen, spread over 4 years, assuming continuous cropping (for example, continuous wheat with take-all). (b) Airborne polycyclic pathogen spread over one growing season for a winter-sown cereal.

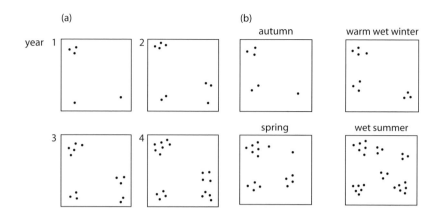

disease outbreaks and disease progress throughout a cropping region. A full description of these models is beyond the scope of this chapter, but a more detailed account can be found in Agrios (2005), Gilligan (2008), and Robbins et al. (2006).

Patterns of Epidemics

The spatial pattern of disease in a crop is therefore related to the life cycle of the pathogen, the prevailing climate and weather patterns (which in the twenty-first century are becoming increasingly erratic), and the crop rotation. For example, if a disease outbreak is of a monocyclic nature (for example, as is the case with soilborne pathogens), disease epidemics are often aggregated into widely dispersed patches within a crop (**Figure 5.11**) as plants growing near the source of the soilborne inoculum, such as overwintering hyphae, become infected. The disease then slowly spreads through successive crops, with the patches enlarging over time. However, unlike polycyclic diseases, patch enlargement is slow and can be checked by the application of crop rotations and other disease control methods (see Chapter 10).

The extent of growth of polycyclic pathogens in a crop is quite different to that of monocyclic pathogens. Polycyclic pathogens typically have several cycles of regeneration within one growing season, and can quickly spread out from the initial outbreak. Using *Septoria tritici* on wheat as an example, an initial outbreak may occur in one or two patches in a field, particularly where local microclimate favors the early growth and reproduction of the pathogen. However, it can spread rapidly across the crop when seasonal climate favors spore dispersal, and in the case of *S. tritici*, spores are dispersed by wind and can be readily blown onto neighboring plants. This process, if allowed to continue unchecked, will eventually result in the collapse of the crop.

An important controlling element for the above scenarios is the disease gradient (**Figure 5.12**). This is the distance of the host from a source of

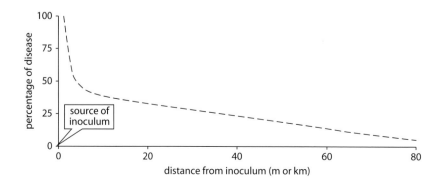

Figure 5.12 Disease gradient. The greater the distance of a crop from the focal point of a disease inoculum, the lower the probability of infection.

inoculum, and it is typically shorter for soilborne pathogens than for pathogens that utilize airborne propagules. Using the disease gradient as a reference to controlling a disease outbreak, growers will diversify their choice of cropping systems and crop varieties in order to increase the distance from the inoculum source and thus maximize the disease gradient.

This type of disease control forms the backbone of the variety diversification program used by many cereal growers as they try to maximize the life of a new wheat variety or reduce the rate of disease dispersal.

5.4 Summary

In this chapter our aim has been to describe the biological processes that underpin the discipline of epidemiology, rather than to focus on the mathematical modeling of an epidemic. Using specific examples, the chapter has explored the key environmental variables, such as temperature, leaf wetness, relative humidity, and light. The examples provided illustrate how pathogens have evolved to cope with these climatic conditions, which of course correspond to the growth of plants and modern cropping systems, and subsequently how disease outbreaks can rapidly develop into widespread epidemics. The chapter has also used a range of examples to explore the relationship between pathogen life cycles, cropping systems, and subsequent disease outbreaks. Finally, we have explored some simple mathematical models used by epidemiologists in an attempt to describe the spatial and temporal dynamics of plant disease, and we have used a simple model to explore the impact of disease sanitation on the progress of a model epidemic.

Further Reading

Aegerter BJ, Nuñez JJ & Davis RM (2003) Environmental factors affecting rose downy mildew and development of a forecasting model for a nursery production system. *Plant Dis* 87:732–738.

Agrios GN (2005) Plant disease epidemiology. In Plant Pathology, 5th ed, pp 265–291. Elsevier Academic Press.

Anon. (1987) Powdery Mildews of Ornamentals. Report on Plant Disease No. 617. Department of Crop Sciences, University of Illinois Extension Service. http://web.extension.illinois.edu/forestry/publications/pdf/forest_health/UIUC_Powdery_Mildew_of_Lilac.pdf

Anon. (1988) Powdery Mildew of Roses. Report on Plant Disease No. 611. Department of Crop Sciences, University of Illinois Extension Service. https://ipm.illinois.edu/diseases/rpds/611.pdf

Canessa P, Schumacher J, Montserrat AH et al. (2013) Assessing the effects of light on differentiation and virulence of the plant pathogen *Botrytis cinerea*: characterization of the *white collar* complex. *PLoS One* 8:e84223.

Carroll JE & Wilcox WF (2003) Effects of humidity on the development of grapevine powdery mildew. *Phytopathology* 93:1137–1144.

Douglas SM (2008) Powdery and Downy Mildews on Greenhouse Crops. Department of Plant Pathology and Ecology, The Connecticut Agricultural Experiment Station. www.ct.gov/caes/lib/caes/documents/publications/fact_sheets/plant_pathology_and_ecology/powdery_and_downy_mildews_on_greenhouse_crops.pdf

Ficke A, Gadoury DM & Semm RC (2002) Ontogenic resistance and plant disease management: a case study of grape powdery mildew. *Phytopathology* 92:671–675.

Gilligan CA (2008) Sustainable agriculture and plant diseases: an epidemiological perspective. *Philos Trans R Soc Lond B Biol Sci* 363:741–759.

Jacob D, Rav David D, Sztjenberg A & Elad Y (2008) Conditions for development of powdery mildew of tomato caused by *Oidium neolycopersici*. *Phytopathology* 98:270–281.

Jones DG (1987) Plant Pathology: Principles and Practice, pp 89–104, Plant Disease Epidemiology. Open University Press.

Jung T, Hudler GW, Jensen-Tracy SL et al. (2005) Involvement of *Phytophthora* species in the decline of European beech in Europe and the USA. *Mycologist* 19 issue 4: 159–166.

Kranz J & Royle DJ (1978) Perspectives in mathematical modelling of plant disease epidemics. In Plant Disease Epidemiology (Scott PR & Bainbridge A, eds), pp 111–120. Blackwell Scientific Publications.

Oudemans P and Coffey D (1988). Phytophthora citricola: Advances in our Understanding of the Disease pgs 16–24. A report to the Avocado Research Advisory Committee, University of California, Riverside

Robbins M, Schimmelpfennig D Ashley E et al. (2006) The Value of Plant Disease Early-Warning Systems: A Case Study of USDA's Soybean Rust Coordinated Framework. Economic Research Report No. 18. Economic Research Service, US Department of Agriculture.

Sansford C, Inman A & Webber J (2010) Development of a pest risk analysis for *Phytophthora ramorum* for the European Union: the key deliverable from the EU-funded project RAPRA. In Proceedings of the Sudden Oak Death Fourth Science Symposium. US Department of Agriculture, Forest Service.

Van der Plank JE (1960) Plant Pathology: Analysis of Epidemics. Academic Press.

Van der Plank JE (1963) Plant Diseases: Epidemics and Control. Academic Press.

Vettraino AM, Natili G, Anselmi N & Vannini A (2001) Recovery and pathogenicity of *Phytophthora* species associated with a resurgence of ink disease in *Castanea sativa* in Italy. *Plant Pathol* 50:90–96.

Waggoner PE & Horsfall JG (1969) EPIDEM, a simulator of plant disease written for a computer. *Conn Agric Exp Stn Bull* 698:80.

Section 2
Compendium of Plant Diseases

The compendium of plant diseases will focus on organisms that cause economically important diseases in a range of different crop and ornamental plants.

Chapter 6: Fungal Diseases

Chapter 7: Bacteria Diseases

Chapter 8: Viral Diseases

Chapter 9: Other Diseases

Chapter 6
Fungal Diseases

Fungi were once considered to be primitive non-photosynthetic plants, probably because they have what superficially appear to be similar structures (for example, stems, roots, etc.), and also because most multicellular species are sedentary and have similar requirements for life (for example, growth media, water, minerals and other nutrients). Moreover, wherever plants are found, fungi are usually not too far away. Indeed many plants, especially forest trees, are dependent on their relationship with fungi for survival.

Interactions between plants and fungi are believed to go back to the origins of the two kingdoms, the co-evolution of which has therefore played a substantial role in the development of both groups up to the present day. These interactions are highly complex and take many forms. Saprophytic and mutualistic relationships are by far the most common, but despite this, the fungi have by far the largest number of representatives of plant pathogens compared with other kingdoms. There are incalculable numbers of fungal pathogens of plants; indeed, a plant can express multiple disease symptoms due to infection by one or more disease organisms (Figure 6.1). Although the fungi are taxonomically complex, it is possible to group these organisms together, based on such criteria as genera, reproductive cycle, conditions that promote pathogenesis in the host, and pathogenic effects on the host. This grouping of pathogenic organisms aids the pathologist and/or agronomist in determining the cause of disease and thus in advising people on the most appropriate treatment strategy to adopt.

The impact of fungal pathogens on plant health and food security is still fairly significant, accounting for 10–15% of all crop losses during a production year, and of course the fungal pathogens are detrimental to ornamental gardens, nurseries, and numerous parks and gardens around the world. The major fungal groups that cause plant disease belong to the basidiomycetes, the ascomycetes, and the deuteromycetes. The latter are more commonly known as the imperfect fungi, due to the lack of a sexually reproductive phase in the **Deuteromycota**. Many of these fungal pathogens infect key crops such as the cereals (for example, wheat, rice, and barley), oilseed rape, and tuber crops (for example, potato). In addition, these pathogens are a significant problem for fruit growers, infecting fruit crops such as apples, pears, and soft fruit (for example, strawberries), and also causing major problems for vegetable growers. It is impossible to cover all the fungal pathogens in a compendium of plant diseases within an introductory textbook on plant pathogens. For this reason we have chosen to highlight a selection of specimens that are of economic significance and/or of interest to ornamental growers. Diagnostic features and, where possible, a range of microscopic characteristics of the disease organism are illustrated.

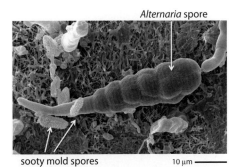

Alternaria spore

sooty mold spores   10 μm

Figure 6.1 An example of a plant that has more than one pathogen infection at the same time. Here an ear of wheat is infected with a sooty mold, but note also the distinctive shape of the large *Alternaria* spore that dominates the image.

(a)

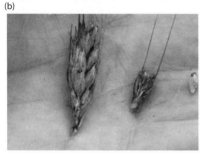

(b)

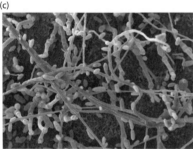

(c)

Figure 6.3 *Alternaria* **infection of brassica.** The arrow indicates the concentric rings that are a diagnostic characteristic of this disease.

Figure 6.2 (a) Typical sooty mold symptoms on ears of wheat, (b) sooty mold on barley, with black spores released onto the field worker's hand, and (c) hyphae and conidia of sooty mold on ears of wheat.

6.1 *Alternaria brassicicola*

Family: Pleosporales

Genus: *Alternaria*

Common name: Alternaria (in cereals, known as sooty molds)

Main hosts: Numerous hosts (over 380 species), including the cereals, in which *Alternaria* is one of many fungal species that cause sooty molds of the ear (**Figure 6.2**). The fungus also infects onions, tomatoes, and ornamental crops such as *Agave* species, and of course brassica crops (**Figure 6.3**).

Life Cycle: The fungus has a dark-colored mycelium that ages to produce short erect conidiophores that have either single or branched chains of conidia. The conidia of *Alternaria* species are large, pear-shaped, and multicellular (Figure 6.1), and they overwinter on plant debris. This fungus can also overwinter as a mycelium. The conidia are easily detached from the mycelium and conidiophores, and are then carried by wind and rain splash, alighting on fresh plant material to start a new cycle of infection. *Alternaria* species are mostly saprophytic and hence infect dead plant tissue, including stems, leaves, and flowers. In addition, because these species are necrotrophic, they leave a distinctive concentric ring that radiates out from the point of infection. The spores of this fungus are constantly present in the atmosphere, and are one of the common causes of human allergies, which can develop into a fatal asthma. Furthermore, *Alternaria* species produce a range of toxins (30 toxins have been isolated to date), such as alternariol, a compound with antifungal and phytotoxic activity.

Treatment: Control is commonly achieved by using disease-resistant cultivars and seed treatments (see Chapter 10), which should be coupled with good biosecurity and sanitation protocols, such as cleaning up and burning dead plant material. Crop rotations will also help to reduce sources of in-field inoculum. In glasshouse cropping systems, covering the glass with a UV light-absorbing film will help to reduce the incidence of *Alternaria* infections on crops, as filtering out UV light inhibits the germination of *Alternaria*. Chemical treatments are available based on compounds such as the strobilurins and the triazoles, including mixtures such as trifloxystrobin and tebuconazole.

6.2 *Aspergillus niger*

Family: Trichocomaceae

Genus: *Aspergillus*

Common name: Black mold

Main hosts: Members of the Solanaceae, Vitaceae, Amaryllidaceae, and Fabaceae

Grapes and grape-derived products are of economic importance worldwide. Economic losses resulting from pathogenic fungi are significant because these organisms can induce diseases and contaminate products with mycotoxins. *Aspergillus niger* is one of the most common species of the fungal genus *Aspergillus*. It is a common contaminant that causes black mold disease on many fruits and vegetables, including grapes, onions, and peanuts. Some species of *Aspergillus* section *Nigri* can produce ochratoxin A, a mycotoxin that contaminates grapes and their derivatives, such as wine. Colonies of these fungi are present on the berry skin from fruit setting, and they increase in quantity from the early stages of ripening through to harvest, with a peak at ripening. The occurrence of *A. niger*-infected fruits is closely associated with climatic conditions during this ripening stage and with the geographical location, and is further responsible for significant losses due to post-harvest decay.

Life Cycle: Molds produce asexual reproductive structures called spores, which have an important role in dispersal, being carried by wind as well as by foraging insects. Under appropriate conditions of humidity and temperature, germination occurs, producing mycelia that reproduce and eventually sporulate again, spreading the infection as a result.

Treatment: *A. niger* can be treated with antibiotics, chemicals, and antibiosis. However, biological control is the best and most effective treatment. The antifungal activity of perillaldehyde (PAE) has promising potential for post-harvest control of grape spoilage. Ultraviolet-C is also a promising emerging technology for reducing the potential ochratoxigenic risk in grapes.

Note: Another species of the genus *Aspergillus*, namely *A. oryzae*, which infects rice, is fermented and used in the making of soy sauce.

> **6.1 Tables Turned on Fungal Disease of Rice**
>
> *Aspergillus oryzae*, the disease organism that infects rice, is fermented and used in the manufacture of soy sauce, a well-known flavoring that has been used in oriental cuisine for hundreds of years.

6.3 *Blumeria graminis* f. sp. *tritici*

Family: Erysiphaceae
Genus: *Blumeria*
Common name: Powdery mildew of wheat
Main hosts: Wheat

The Erysiphaceae is the only family in the order Erysiphales. However, it consists of a highly diverse range of genera and species, and causes a disease commonly known as powdery mildew in an equally diverse range of hosts, although typically it only infects higher plants such as sweet pea, courgettes, oak, and clematis (**Figure 6.4**), to name just a few. An important example is *Blumeria graminis*, which infects grass species. This includes *Blumeria gramini* f. sp. *tritici*, which is specific to wheat (a global staple), and is therefore of significant economic importance.

Powdery mildew is the most widespread and commonly observed cereal disease across all the cereal-growing regions of the world. Symptoms occur

Figure 6.4 Powdery mildew on (a) sweet pea leaf, (b) courgette leaf, (c) oak leaf, and (d) clematis bloom. A color version of this figure can be found in the color plate section at the end of the book.

Figure 6.5 Powdery mildew: (a) wheat leaf showing characteristic white fluffy pustules (indicated by arrow), (b) conidia and mycelia erupting from the host tissue, and (c) the black resting spores, known as cleistothecia (indicated by arrow).

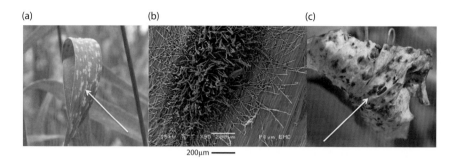

on all aerial parts of the plant, first appearing as small chlorotic flecks on the leaves and stems, shortly followed by patches of white, fluffy pustules (Figure 6.5a) which produce thousands of conidia (Figure 6.5b). As the disease advances, black resting spores (known as cleistothecia) are embedded in the mildew pustules (Figure 6.5c).

Life Cycle: In Europe, the fungus overwinters as mycelium in host tissue such as volunteer plants which have regrown from seed that spilt from the combine during harvest. Cleistothecia can also aid the survival of powdery mildew for several weeks in the absence of host material. Infection of autumn-sown cereals arises from wind-blown conidia, which have the ability to germinate across a wide range of temperatures (5–30°C), with the optimal temperature range being 15–20°C and optimal relative humidity over 90%. Under optimal conditions the latent period is about 7 days. Ascospores arising from the cleistothecia are also sources of inoculum, which accelerates crop infection rates. The pathogen is restricted to the upper cell layers and is an obligate parasite, but it seldom kills the plant, due to its biotrophic characteristics.

Treatment: Cultural control is difficult and relies on mildew-resistant cultivars and reduced planting density in organic systems. The disease is more widespread in crops that are grown in nutrient-rich soils, as this results in lush plants with soft susceptible epidermal tissue. Chemical treatment with a triazole-based fungicide provides good protection, and this may be formulated with a succinate dehydrogenase inhibitor (SDHI) fungicide such as bixafen to provide a curative treatment in regions where there is a high mildew pressure.

6.4 *Botrytis cinerea*

Family: Sclerotiniaceae
Genus: *Botrytis*
Common name: Gray mold
Main hosts: Many flowers, fruits, and vegetables

Gray molds caused by the fungus *Botrytis cinerea* are one of the most important diseases of many crops, including brassicas (Figure 6.6), and ornamental plants (Figure 6.7) worldwide. This organism has a significant impact on soft fruit such as strawberries. The disease is a problem not only in the field, but also during storage, transit, and marketing of the strawberries, due to the onset of severe rot as the fruits begin to ripen.

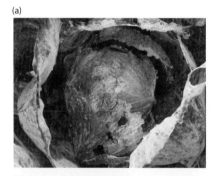

Figure 6.6 Gray mold (*Botrytis cinerea*) on red cabbage (*Brassica oleracea* var. *capitata* f. *rubra*). (a) Fungus infecting the heart during early winter, speeding up tissue decay. (b) A mass of hyphae and conidia erupting from the leaf tissue of infected red cabbage.

Life Cycle: The fungus forms dark hardened structures (sclerotia) in dead plant tissue, and these can persist in the soil for years. The sclerotia germinate to produce the sexual structures, which in turn release a second, sexual spore. These sexual spores can initiate infections, but most of the damage is caused by spread of the airborne, asexual spores (conidia) released from the fuzzy gray fungal growth. As the infected strawberry leaf begins to die, the pathogen enters an active stage, colonizing the leaf and obtaining its nutrients from the dead tissue. Spores then form, and when environmental conditions are appropriate (temperatures in the range of 18–24°C during damp or rainy weather) they are dispersed by water splash and/or wind onto newly emerging leaves or blossoms. Immature fruits become infected primarily through blossom infections. Once the berries begin to ripen, the fungus is able to colonize them and sporulate, producing the mold that is often seen in the field.

Treatment: None of the commercially available strawberry varieties have any degree of resistance to gray mold, but the disease can be controlled by both chemical and non-chemical measures. Non-chemical practices include strict hygiene (especially under glass), prompt removal of dead and dying leaves, buds, and flowers, reduction of humidity by improving ventilation, and avoiding overcrowding of plants. Chemical methods include the use of fungicides, which are critical in problem fields during early and full bloom. These fungicides are targeted to limit flower infection that leads to fruit infection, and should limit the need for late-season applications to the fruit. In areas prone to dense fog and in environmental conditions that are favorable to disease development (that is, temperate conditions), use of a protective fungicide to prevent germination of spores is expedient. Thereafter, regular spraying schedules can be applied according to disease pressure and environmental conditions.

Figure 6.7 Gray mold infestation of a rose flower, showing a mass of white mycelium on a decaying flower bud.

6.5 *Botrytis fabae*

Family: Sclerotiniaceae
Genus: *Botrytis*
Common name: Chocolate spot
Main hosts: Broad bean (*Vicia faba*) and common vetch (*Vicia sativa*)

Chocolate spot is the most common disease of broad bean plants, affecting the flowers, leaves, and bean pods (Figure 6.8). The broad bean is the only commercial host for *Botrytis fabae*, although this disease is not to be confused with another similar disease, caused by *Botrytis cinerea*, which is also sometimes called chocolate spot. *B. cinerea* has a broader range of hosts, although the broad bean itself is less susceptible to this species. In

(a)   (b)

Figure 6.8 Chocolate spot (*Botrytis fabae*), the most common disease of broad bean (*Vicia faba*), with the characteristic spotting on (a) the leaves and (b) the seed pods.

Figure 6.9 If chocolate spot in broad beans is allowed to progress, the infection may reach the stems, causing complete collapse of the plant.

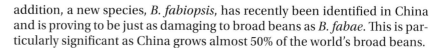

addition, a new species, *B. fabiopsis*, has recently been identified in China and is proving to be just as damaging to broad beans as *B. fabae*. This is particularly significant as China grows almost 50% of the world's broad beans.

B. fabae favors humid but cool conditions, and is therefore ideally adapted for survival in the temperate winter or early spring conditions that are typical for the broad bean growing season. Where these conditions persistently prevail, stems may also become infected and collapse of the whole plant may result (**Figure 6.9**). Yield losses can be significant during severe infection episodes, although milder infections, where conditions are warmer and drier, have little effect on yields, despite clear visual evidence of infection.

Life Cycle: This disease is transmitted as conidia, and passes quickly from plant to plant via wind or water splashing. Brown necrotic spots result from infection as the conidia germinate and penetrate the host tissue. Developing conidiophores grow on the dead tissue, producing ever more conidia. At the end of the growing season, the disease persists in the soil and in the remaining plant debris as sclerotia, or it may sometimes infect common vetch as an alternative host, when this is available, increasing sclerotia numbers. When the new season begins, the sclerotia produce mycelia that in turn produce conidiophores.

Treatment: As high humidity favors this disease, decreasing planting density may reduce the humidity by increasing airflow around the plants, although this will also have implications for final yields. Removal of vetch from the fields may reduce the numbers of sclerotia that enter the dormant phase between growing seasons. Fungicide sprays can be quite effective, especially if used during flowering, when the fungus is in the early stages of growth. To date, no resistant varieties have been cultivated. However, recent research indicates that fungal antagonists from the genus *Trichoderma* may be useful for developing biological treatment regimes.

6.6 *Choanephora cucurbitarum*

Family: Choanephoraceae
Genus: *Choanephora*
Common names: Cucurbit stem rot, cucurbit blossom rot, wet rot
Main hosts: Cucurbits, peppers, okra, pulses

Choanephora cucurbitarum is a fungal disease that affects mainly cucurbits (courgettes, squash, etc.) (**Figure 6.10**), but can also infect chillies and some other peppers, as well as okra and some of the pulse groups. This disease causes blossom blight if infection occurs early in the season, or rotting of the fruits if infection takes hold later in the year. It is a common disease that thrives in conditions of mid to high temperatures and high humidity, and is therefore typically found in warmer countries, or in glasshouses in temperate regions. As this disease affects so many commonly grown fruits globally, it is economically significant.

Life Cycle: *Choanophora* species prefer warm, humid conditions. However, when environmental conditions do not meet these temperature and humidity requirements, this fungus can survive for long periods in the form of dormant spores. If the environmental conditions do favor growth but no suitable host is available, the fungus can switch lifestyles and live saprophytically on dead and decaying plant material. Once the host plants grow and produce flowers, the fungus switches back to a pathogenic lifestyle, and infectious spores quickly infect the new flowers and then spread to the fruits.

Figure 6.10 Yellow courgette variety with severe cucurbit fruit rot (*Choanephora cucurbitarum*). Note the large branching spore-bearing heads. (Courtesy of David B Langston, University of Georgia.)

(a)   (b)

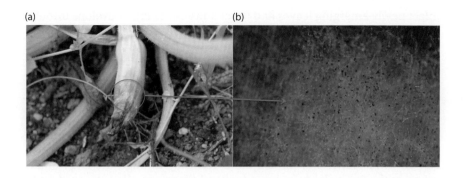

Figure 6.11 **Yellow courgette variety with fungal soft rot (*Rhizopus stolonifer*), a pathogen suited to cooler conditions.** (a) Note the similarity of symptoms on the fruits to *Choanophora* infection as the disease works its way from the flower up along the body of the fruit. However, in this case (b) the spore-bearing heads are not branching.

Treatment: Disease management is complex and time-consuming, as flowering is not coordinated, with new flowers opening daily and presenting new host material for the infectious spores. Some cultural control measures, such as increased plant spacing and good air movement and adequate drainage in glasshouses, may reduce infection.

Note: This disease can be easily confused with *Rhizopus stolonifer* (fungal soft rot, also known as black bread mold, or mucor), another fungal disease that is similar in appearance and affects many of the same hosts, but is tolerant of lower temperatures (**Figure 6.11**).

6.7 *Cylindrosporium concentricum* Grev. and *Pyrenopeziza brassicae*

Family: Dermateaceae

Genus: *Cylindrosporium* and *Pyrenopeziza*

Common name: Light leaf spot

Main hosts: Brassicas, in particular oilseed rape, but also vegetable brassicas and wild brassica plants, and volunteer oilseed rape

Both of these organisms cause a disease commonly known as light leaf spot, which affects all brassicas. However, for arable farmers light leaf spot is a major problem in oilseed rape, with crop losses estimated to be in the region of £100 million in the UK in 2010. *Cylindrosporium concentricum* is the asexual stage of light leaf spot, and *Pyrenopeziza brassicae* is the sexual stage, once rare in the UK, but now more common in England. Light leaf spot infects leaves, stems, flower buds, and seed pods. The early symptoms are chlorotic to necrotic lesions on the margins of leaves, which are surrounded by white spores; these lesions coalesce over time to form larger necrotic patches. Infected leaves have a slower growth rate at and around the infection site, which results in distorted leaves. Elongated lesions can also be detected on the stems of infected plants.

Life Cycle: The pathogen is favored by cool temperatures (optimum range 5–15°C) and wet conditions. It overwinters on crop debris as a mycelium, and can also overwinter on other field crops belonging to the family Brassicaceae. Field debris are initially infected with conidia from the standing crop in late summer and early autumn, and once the fungus is on the crop debris it enters a sexual phase and develops apothecia, which are typically less than 1 mm in diameter. Ascospores develop in the apothecia and are then released by wind and blown onto emerging winter-sown oilseed rape. Ascospores can be wind-blown for many miles, thus infecting crops in the neighboring fields within a crop rotation. After this initial infection, the disease spreads by rain-splashed dispersal of ascospores from the initial infection site, and thus the disease appears patchy in the field. As winter-based infections often originate at the tips of leaves, the pathogen is able to infect newly emerging leaves and floral buds as the crop develops through the winter and early spring season. On infected volunteer oilseed rape plants present in the crop and/or in the field margins, and on wild brassica relatives in the field margins, the pathogen produces conidiophores that are able to spread into the crop, but this is a very much localized dispersal of inoculum.

During stem extension (in the summer) the pathogen enters another sexual phase and produces more apothecia on senesced leaves lower down in the crop canopy. Once again the apothecia release ascospores, which can be wind-blown into the crop and vertically through the crop, finally infecting the flower buds and mature floral tissue as well as the seed pods. Simultaneously, the pathogen releases conidia from wild brassica relatives growing in the field margins.

Treatment: The treatment strategy is similar to that for phoma stem canker, and for commercial arable crops there are a range of fungicides available, with maximum application and dose recommendations to avoid the build-up of fungicide resistance in the target pathogen.

6.8 Diplocarpon rosae

Family: Dermateaceae
Genus: *Diplocarpon*
Common name: Black spot
Main hosts: Cultivated roses, and wild roses such as *Rosa canina*

Life Cycle: Conidia are produced in the black spots (Figure 6.12) and are released via rain splash, handling, and/or pruning of the plant. The black spots can overwinter and hence the conidia produced in these overwintering spots are a source of new-season inoculum. The disease can also overwinter in diseased canes and on fallen leaves. The fungus is active in warm moist regions, and is problematic in the south-west of the UK, where severe infections cause complete leaf drop, reduced vigor, and aborted buds.

Treatment: Treatment using chemicals is now difficult in Europe because of a widespread review of pesticides, which has resulted in the withdrawal of many active ingredients. New compounds based on myclobutanil are available. However, good cultural practice such as collecting up and burning of infected fallen leaves, disinfecting pruning shears, and washing hands after handling infected roses will help to reduce inoculum levels. Pruning out diseased canes from a bush or climbing rose will also contribute to overall health of the rose garden. Finally, there are a number of resistant cultivars, but over time many of these cultivars succumb to black spot as the pathogen evolves new races that negate the resistance mechanism.

Figure 6.12 (a) Black patches of rose black spot on foliage, and (b) conidia of rose black spot erupting from black patches on the leaf.

(a)   (b)   (c)

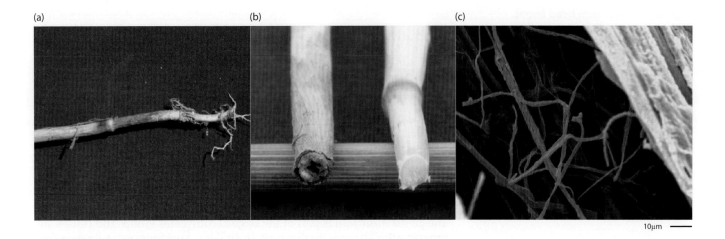

10μm ——

6.9 *Fusarium culmorum* and *Microdochium nivale*

Family: Nectriaceae

Genus: *Fusarium* and *Microdochium*

Common name: Foot rot

Main hosts: Wheat, barley, and many other host species

Seedling blight and foot rot (Figure 6.13) are common problems for cereal growers, and in this situation the disease does not cause wilt but results in the rotting of the stem and/or the seedlings (another form of damping off). Rotting of the roots, lower stems, and crowns of plants results in white-heads in the case of cereals. The causal organisms are not solely confined to the cereals, and are known to infect numerous vegetable crops, bulbs, and tubers.

Life Cycle: Both species survive saprophytically in soil and plant debris. *Fusarium culmorum* infections occur in autumn-sown crops, and the disease is favored by warmer drier soils, with progress into the crop occurring within the optimum temperature range of 10–15°C. The stem-based infections give rise to perithecia under moist conditions in the spring months. The perithecia release ascospores that infect the flag leaf. *Microdochium nivale* arises from macroconidia that are formed on crop debris, and this organism prefers slightly warmer temperatures, in the range of 10–20°C.

Treatment: Sowing treated seed will enhance crop establishment. Seed treatments are typically based on a triazole, such as prothioconazole, and subsequent tank mixes of *Fusarium* active fungicides help to protect the crop during the growing seasons. This should be combined with good crop husbandry, such as rotating soil cultivation techniques from minimal tillage through to conventional ploughing methods that help to bury plant debris from the previous crop.

6.10 *Fusarium graminearum* and other *Fusarium* species

Family: Nectriaceae

Genus: *Fusarium*

Common name: Fusarium head blight (FHB)

Main hosts: Wheat and barley

Five main causal organisms cause head blight (Figure 6.14), namely *Fusarium avenaceum* (*Gibberella avenacea*), *F. culmorum*, *F. graminearum*

Figure 6.13 *Fusarium* **foot rot.** (a) Symptoms on mature plant. Note the stunted root growth. (b) The effect of *Fusarium* foot rot on leaf sheaths and stems. Compare the dead material at the base of the sample on the left with the healthy leaf sheaths and stem on the right. Also note the presence of black pycnidia on the infected stem on the left. (c) Hyphae of the causal organism in a section of wheat stem.

Figure 6.14 *Fusarium* **head blight.** (a) Symptoms on the ear. Note the spikelets adjacent to the field worker's thumb, and the presence of pycnidia. (b) The mass of hyphae and the development of conidia.

(a)   (b)

(*Gibberella zeae*), *F. poae*, and *Michrodochium nivale* (formerly *F. nivale*). The genus *Fusarium* is part of the large ascomycetes group, and causes significant crop losses in both wheat and barley. Furthermore, *Fusarium* species are known to produce mycotoxins, in particular *F. graminearum*, which produces deoxynivalenol (DON). These mycotoxins are not broken down during cooking, and thus are quite harmful to humans, as well as to animals fed on contaminated grain. The main symptoms of mycotoxin poisoning in humans are nausea, fever, headache, and vomiting. It is believed that mycotoxins are part of the mechanisms that the fungus uses to infect plants, as these mycotoxins are most likely to break down plant defense mechanisms. The summer of 2012 in the UK was exceptionally wet, and cereal farmers struggled to control FHB in both wheat and barley crops. The final yield loss to FHB for UK cereal growers in 2012 was estimated to be around 25%. In the USA, between 1998 and 2000 financial losses due to FHB were estimated to be in the region of US$ 2.7 billion, following a protracted FHB epidemic.

Life Cycle: The fungus can survive saprophytically in the soil or on crop debris, and also on a range of different plants that overwinter in the stubble. FHB can also be seed-borne on the cereals, and this demonstrates the need to use certified seed (see Chapter 10). The development of the disease in the cereal heads occurs under conditions of high summer temperatures and high relative humidity during the heading phase of the crop life cycle. Once the causal organism has established in a cereal spikelet, it appears water soaked and eventually dries out to a pinkish straw-like color (Figure 6.14a). The pink coloration is an indication of growth of mycelium, from which abundant conidia are produced, which subsequently infect adjacent spikelets and plants. As the kernels in the spikelets mature they take on a light-brown shriveled appearance due to the continued growth of the mycelium, which erupts from the pericarp.

Treatment: Control is quite difficult, as there is poor resistance in most cereal varieties, both in Europe and in the USA. This is because resistance mechanisms are thought to be based on a complex array of resistance genes in the host. The best practice is to use crop rotations, but this has limitations due to the wide host range of the numerous *Fusarium* species. However, breaking the cycle of wheat → barley rotation with a broadleaf crop such as oilseed rape or a legume will help to reduce the levels of *Fusarium* inoculum in the fields. Chemical treatment is possible (see Chapter 10), but reports of mycotoxin production by *Fusarium* species treated with fungicides are not uncommon.

6.11 *Fusarium oxysporum* f. sp. *narcissi*

Family: Nectriaceae

Genus: *Fusarium*

Common name: Narcissus basal rot

Main hosts: *Narcissus* species (daffodils)

Basal rot is one of the most serious diseases of daffodils, and is caused by the soilborne fungus *Fusarium oxysporum* f. sp. *narcissi*. *Fusarium* diseases affect a wide range of bulb crops. Plants infected with *Fusarium* typically die down prematurely, and frequently there is a brown-colored rot of the bulb. The fungus can progress throughout the bulb, thereby killing it.

Life Cycle: This is a warm-weather disease that is most prevalent on acid, sandy soils. The fungus is transferred from the soil to the bulb via the roots, or by adhering to the outside of the bulb when it is lifted. Bulbs can become infected during storage or after planting, and spores become widespread and are viable for over 10 years in the soil. The disease is favored by warm temperatures, although it can develop in cooler conditions. It is also favored by moist soils that are optimal for plant growth. *Fusarium* species reproduce by means of asexual spores, and overwinter either in this form or as mycelia. They can switch between parasitic and saprophytic lifestyles as required.

Treatment: Control is mainly achieved through the use of resistant cultivars. The avoidance of fresh manure or excessive nitrogen is essential, and early lifting of the bulbs is desirable. The lifted bulbs should immediately be sprayed with a suitable fungicide and dried rapidly in a good airflow. They should be stored at a low temperature of 17–18°C, with planting in late September or early October when soil temperatures are lower. Bulbs in storage should be inspected regularly and any that are soft should be destroyed. Hot water treatment can also be used to control the fungus in *Narcissus* bulbs, whereby the bulbs are submerged in water heated to exactly 44.5°C for 3 hours. If the fungus has contaminated an area, this should be kept free of bulb-forming plants for 3 years.

6.12 *Gaeumannomyces graminis* var. *tritici*

Family: Magnaporthaceae

Genus: *Gaeumannomyces*

Common names: Take-all, white head

Main hosts: Cereals (especially wheat and barley) and other grasses

Symptoms of this disease include darkening and poor development of roots (**Figure 6.15**), early maturation, poor overall growth, white seed heads with poor grain fill, and yellowing of leaves.

Life Cycle: Although this fungus does reproduce sexually by means of perithecia (fruiting bodies) that contain asci, which in turn produce long, thin ascospores, it is asexual reproduction that is key to its pathogenic success. *G. graminis* infects the roots of host plants mainly by vegetative growth of hyphae, and is able to overwinter in plant debris prior to infecting new plants the following growing season.

Treatment: Crop rotations can be successful, as the fungus will not survive a whole year without a host. However, this disease can persist in wild relatives of the crop plants, and reinfect when the grower rotates back to cereal. One interesting point about this disease is that it suffers from what has become known as "take-all decline." If left unchallenged, *G. graminis* will lose its virulence naturally after a few years, rendering it largely harmless to the crop.

6.13 *Hymenoscyphus fraxineus* (also known as *Chalara fraxinea*)

Family: Helotiaceae

Genus: *Hymenoscyphus*

Common name: Ash dieback or ash decline

Main hosts: Ash trees (*Fraxinus* species)

Figure 6.15 Darkening and poor root development in wheat infected with *Gaeumannomyces graminis* var. *tritici* (take-all), rendering the plant easy to pull up from the soil.

The fungus *Hymenoscyphus fraxineus* causes ash dieback—a disease that has been much in the news in recent years. This fungus attacks the leaves and crown of ash trees, and grows through the inside of the tree, killing all the tissue in its path and preventing other parts of the tree from accessing water and nutrients. Infected trees do not recover from infection and are often killed, with young trees being the most vulnerable. *H. fraxineus* causes infection from June to October, and is most prevalent during July and August, especially in moist, humid conditions. The fungus was first identified in the UK in trees imported to a nursery in Buckinghamshire in 2012, and was confined to central and eastern Britain in 2012, but has now spread throughout most parts of the UK.

Life Cycle: The infectious sexual spores of the fungus are produced by fruiting bodies (apothecia) and can be wind-blown over long distances (20–30 km). The apothecia are produced from June to October on the ash leaf petiole and rachis. The spores land on leaves or other parts of the trees. From the leaf, the fungus makes its way down the petiole, rachis, and stem. The fungus can also produce asexual spores, but these are believed to be non-infectious and can only be spread over short distances by water splash. This pathogen overwinters in the leaf litter.

Treatment: This is a notifiable disease in the UK, so suspected infections have to be reported to the Forestry Commission. There is no known chemical control at the time of writing. All infected trees must be completely removed and a *cordon sanitaire* has to be established in the infected region. Mature trees are less susceptible to the disease, and ongoing research is evaluating the genetic diversity of both the host and the pathogen in order to inform future disease-resistant breeding programs.

6.14 *Leptosphaeria maculans* and *L. biglobosa*

Family: Leptosphaeriaceae
Genus: *Leptosphaeria*
Common name: Phoma stem canker or blackleg
Main hosts: Brassicas, especially oilseed rape

The initial symptoms of this disease can be similar to those of light leaf spot of oilseed rape (that is, small leaf lesions with white borders), but the lack of pycnidia (see below) in light leaf spot disease symptoms is a diagnostic feature, as is the presence of pycnidia, which indicates that the disease is *L. maculans*. To the novice, the literature on both phoma stem canker and light leaf spot is not entirely clear, and care should be exercised when identifying the causal organism.

Life Cycle: This pathogen can be spread by wind-blown ascospores and/or rain splash of conidia in newly planted crops. The first symptoms appear as small circles of black pycnidia on leaves, often with a white to gray halo, as early as September in an autumn-sown crop. The fungus systemically colonizes the plant during what is known as the endophytic stage, and during the systemic invasion the pathogen spreads rapidly throughout the plant via the vascular tissue. It spreads down the main stalk and into the vascular tissues of the leaf stalks, and can invade the cortex cells around the xylem. The pathogen will spread onto seed pods, thus contaminating the seeds within the pod. At the end of the season the fungus causes stem cankers, and is now in a necrotrophic phase, living off dead plant material. The pathogen then overwinters as a saprophyte on crop debris. The main symptoms on leaf foliage in autumn are caused by *L. maculans*. Then secondary symptoms appear on the stems during the summer, and these are in fact caused by another species, known as *L. biglobosa*, which produces dark lesions and fruiting bodies on the stems of the plant.

Treatment: Phoma stem canker is a major problem for oilseed rape growers in the UK, and in 2010 Defra estimated that crop losses due to phoma stem canker were in the region of £80 million. The disease control strategy consists of a combination of fungicides and sowing of disease-resistant cultivars. However, there are increasing problems with cultivars that are resistant to phoma stem canker in that there is mounting evidence that the major R-gene resistance mechanism (see Chapter 10) is breaking down in certain countries around the world. Cultivation techniques such as deep ploughing of stubble will help to reduce inoculum levels. Then timely applications of triazole-based fungicides, such as prothioconazole, will help to keep the crop clean, but the use of fungicides needs to be managed to avoid fungicide resistance appearing and/or dominating the contemporary races of phoma stem canker. Management of light leaf spot is similar to that of phoma stem canker.

6.15 *Magnaporthe oryzae* (formerly called *M. grisea*)

Family: Magnaporthaceae

Genus: *Magnaporthe*

Common name: Rice blast

Main hosts: Rice and the following strains of rice: M-201 (which is the most vulnerable), M-202, M-205, M-103, M-104, S-210, L-204, and Calmochi-101

Life Cycle: This pathogen infects as a spore and produces small white to gray spots or lesions, which are elliptical or spindle-shaped, on the culms, leaf collar, leaves, and panicle of rice. The lesions enlarge over time, completely engulfing the leaf or griddling the culm and/or leaf collar and thus killing the plant. Conidia germinate on the surface of the plant and, via the appressorium, the pathogen penetrates the epidermis. Hyphae rapidly grow through the internal structures, and eventually develop conidiophores that produce more conidia. The whole cycle can be completed in 8 weeks. Optimal conditions consist of long periods of free moisture (moisture films on plant surfaces), high relative humidity, and temperatures of 25–28°C.

Treatment: The causal organism is developing resistance to chemical treatments, and is also evolving to overcome the disease resistance that has been bred into a range of rice cultivars. An integrated approach is required whereby cultivar mixes are used across a cropping region and fungicide mixes are used to avoid reliance on just one or two active compounds. Good cultural control is also valuable, including burning diseased straw, and sowing seeds into soils covered by water in order to reduce seed-borne transmission, as the fungus cannot survive the anaerobic conditions of rice paddies. Shading should be removed, as it enhances leaf wetness, and disease-free certified seed should always be used.

6.16 *Monilinia fructigena*

Family: Sclerotiniaceae

Genus: *Monilinia*

Common name: Brown rot

Main hosts: Members of the families Rosaceae and Ericaceae

Monilinia fructigena is a fungal disease that affects orchard crops, especially apples, pears, and stone fruits such as plums. It can also spread to berry fruits and grapes. Brown rot can actually be caused by a number of different fungal pathogens (of which the genus *Monilinia* is the most common and the most

Figure 6.16 Apple with brown rot disease (*Monilinia fructigena*), with inset showing detail of fungal clusters.

Figure 6.17 Apple with advanced brown rot disease (*Monilinia fructigena*). Mummification is well under way, although the fruits have not yet fallen from the tree.

well known), and leads to blossom blight if the infection is expressed early in the season, or rotting of the fruits if the infection takes hold later in the year (**Figure 6.16**). At its extreme, fruit mummification results from water loss (**Figure 6.17**), and, if unchecked, the disease can induce **cankers** in the woody tissues of the host plant.

Life Cycle: The fungus is well adapted to all eventualities. In the winter it resides in the mummified fruits that have remained on the tree or in the cankers in the woody tissue as mycelia that later develop into conidia, ready to infect the breaking buds at the start of the new season. However, if the mummified fruits have fallen to the ground, the fungus produces **apothecia** (cup-shaped fruiting bodies) that can emerge through the covering layer of soil. These apothecia contain the asci, which in turn house the developing **ascospores**. As the ascospores mature they are released into the atmosphere by **dehiscence**, which maximizes the likelihood that they will land on a new host.

Treatment: *Monilinia fructigena* is a common disease across most of Europe and parts of southern Asia. Commercially, the disease is controlled using fungicide sprays in series, which together with development of partially resistant cultivars ensures that high cropping levels are maintained. However, as the disease easily spreads throughout wild populations of related plants, eradication is not possible at the present time.

6.17 *Oculimacula acuformis* (R type) and *O. yallundae* (W type) (formerly *Tapesia acuformis* and *T. yallundae*, respectively, syn. *Pseudocercosporella herpotrichoides*)

Family: Dermateaceae

Genus: *Oculimacula*

Common name: Eye spot

Main hosts: Wheat and barley

This is a common disease in cool cereal-growing regions, such as the UK and mainland Europe, New Zealand, and certain states in the USA. The disease causes honey-brown eye-shaped lesions at the base of the straw, and was once called "straw breaker", due to the fact that the fungus develops a mass of mycelium in the cavity of the straw, which eventually results in crop **lodging**. The disease can also produce aborted ears, which turn white in color and are known as whiteheads (**Figure 6.18**). Definitive identification of eyespot

Figure 6.18 Aborted wheat ears, known as whiteheads, following infection with *Oculimacula* species (eyespot).

Figure 6.19 Scanning electron microscopy image of the long, thin asexual conidia of *Oculimacula* species (eyespot).

is complicated by a number of other foot-based diseases, such as *Fusarium* foot root and sharp eyespot.

Life Cycle: Long thin asexual conidia (Figure 6.19) are produced on cereal stubble during the autumn and winter months. These spores are subsequently rain splashed over short distances during cool (5–15°C) moist periods, and then infect new plants. Ascospores are also produced during the growing season, and may also be involved in crop infection. The disease is typically monocyclic (see Chapter 5). The vegetative life cycle of the disease is typically 15 days, but it can take up to 8 weeks for symptoms to occur on the stem. The sexual apothecia are produced at the end of the growing season. Free water is essential for disease infection, and thus the disease is favored by standing water—a common feature in cereal fields during the winter months.

Treatment: Effective control is achieved by using disease-resistant cultivars and applications of a triazole fungicide such as prothioconazole. Introducing oats or another crop species, other than barley, into a rotation will help to reduce crop-based inoculum, and the use of deep ploughing will help to bury infected crop debris.

The disease eventually results in lodging of the crop, but modern short straw stems have reduced the impact of this disease on harvest logistics. Crop losses are small, at around 5%, but severe and uncontrolled epidemics can result in losses of up to 40%.

6.18 *Phoma clematidina*

Family: Didymellaceae
Genus: *Phoma* incertae sedis
Common name: Clematis wilt
Main hosts: Clematis species

The *Phoma* genus of plant pathogens belongs to the ascomycetes, and is taxonomically complicated. There are a large number of plant pathogens within the genus *Phoma* that are pathogenic to a range of plants, including crops

6.2 Powerful Life-Saving Drugs from a Common Plant Pathogen

Phoma species cause numerous diseases in plants, but they also produce a secondary metabolite called squalestatin. Statins are used medicinally to reduce cholesterol levels in patients. There is also newly emerging evidence that squalestatin could be used as an antimalarial drug, as it inhibits carotenoid synthesis in *Plasmodium falciparum*, resulting in the death of this malarial parasite.

and garden vegetables (for example, *P. solani* which infects potatoes, *P. beta* which causes heart rot of sugar beets, and *P. batata* which produces dry rot of sweet potato). Pathogenic species of the genus *Phoma* cause a range of disease symptoms, including pitting of fruits, rotting of stems (for example, blackleg of cabbage seedlings), damping off of seedlings, and wilts of key ornamentals, such as clematis wilt caused by *P. clematidina* (Figure 6.20). The taxonomy of the order Pleosporales (to which *Phoma* species belong) is complex, with nine sections, and the *Phoma* section contains at least 70 identified species. However, the *Phoma* group is described by taxonomists as "incertae sedis," denoting that it is very difficult to place. Phylogenetic studies indicate that the genus *Phoma* is polyphyletic and contains six distinct **clades**, several of which are in the Pleosporales, but the key *Phoma* pathogens are within the Didymellaceae, which is also taxonomically complex, containing a further 18 clusters.

Life Cycle: In the garden or nursery setting this pathogen is mainly spread by rain splash of conidia, which are typically very small, often less than 15 μm in diameter. The conidia develop within a fruiting body known as a pycnidium (plural pycnidia) (Figure 6.20c), which can be formed on any infected part of the plant. The pycnidium matures and releases the conidia in a sticky matrix, which is subsequently dispersed onto healthy tissue. The conidia randomly stick to hairs and infect the leaf tissue directly through the leaf hair (Figure 6.20d). If the infected plant has some innate resistance mechanisms, the newly infected leaf hair may be abscised, but much more commonly this is not the case and infection occurs.

Treatment: The standard practice with clematis wilt is to prune the plant back to the ground and then pack the base with moist, peat-rich soil and apply a soil drench of a growth stimulant such as liquid seaweed extract. This practice typically strengthens new root growth and helps the plant to cope with the disease. It is advisable to remove all dead plant material, as the fungus can develop thick-walled chlamydospores which can overwinter and start a new cycle of infection. Disease-resistant varieties are available, but the fungus is very aggressive in large-flowered varieties and in varieties that have *C. lanuginosa* in their ancestry, and *Clematis viticella* is also quite susceptible. Resistant species and cultivars include *Clematis* 'Avant-Garde', *C. alpina*, *C. montana*, and *C. orientalis*, among others. Currently there are no recommended fungicides for treatment of clematis wilt in the UK.

Figure 6.20 (a) Damping off of seedlings caused by *Phoma* species; (b) clematis wilt caused by *Phoma clematidina*; (c) the pycnidium fruiting body; (d) conidia attached to hairs of the stem of *Clematis* species.

Figure 6.21 *Puccinia* **species (rust) in leeks.** (a) Infection in crop row; (b) the yellow urediniospores; (c) scanning electron microscopy (SEM) image of urediniospores; (d) SEM image of teliospores.

6.19 *Puccinia graminis* f. sp. *tritici* and *P. recondita* f. sp. *tritici*

Family: Pucciniaceae

Genus: *Puccinia*

Common name: Black stem rust and brown rust, respectively

Main hosts: Wheat and barley are the main cereal hosts for these two rust species

There are numerous rust species that affect many different plant species from both the monocotyledons and the eudicotyledons. Examples include rusts of oats (not crown rust) and rye, rusts of numerous wild grass species, in particular Yorkshire fog (*Holcus lanatus*), and rusts of cultivated bulbs such as onions, shallots, and leeks (Figure 6.21). Rusts also infect trees, and are quite vigorous on the willows (*Salix* species) and on poplar trees (*Populus* species). They also occur on ornamentals such as hollyhocks and daisies (Figure 6.22), and on crops such as broad bean (*Vicia faba*).

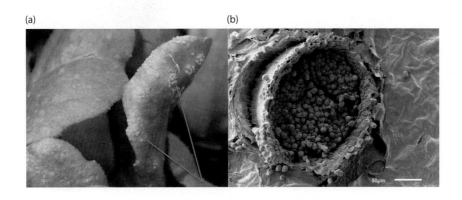

Figure 6.22 (a) Urediniospores of daisy rust and (b) scanning electron microscopy image of a rupturing uredinium packed with mature urediniospores.

(a) (b)

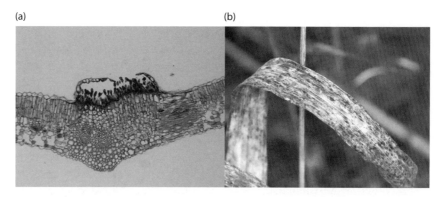

Figure 6.23 (a) **Light microscopy image of** *Puccinia* **teliospores and (b) late-season expression of teliospores on senesced cereal leaf.** (Courtesy of Joanne Knight, Plymouth University.)

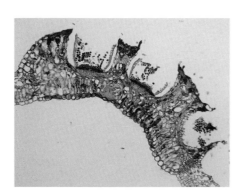

Figure 6.24 Light microscopy image of aecidia with pycnidiospores. (Courtesy of Joanne Knight, Plymouth University.)

Life Cycle: The life cycle of rusts is complicated and involves numerous spore types (Figure 3.11) and resting stages, such as the pycnidia and the overwintering teliospore stage (Figure 6.23). *Puccinia graminis* requires the cereal and a secondary host, namely barberry. In temperate regions this pathogen survives the winter as a teliospore on crop debris. However, the pathogen can survive as a mycelium in mild areas where winter temperatures do not fall below 0°C. The teliospores germinate in the spring, producing a mass of basidiospores that are wind-blown for up to a quarter of a mile, where they will then alight on barberry; in Europe this is the common berberis (*Berberis vulgaris*), and in the USA it is known as American barberry (*Berberis canadensis*). Once a basidiospore has landed on a barberry leaf it quickly germinates and forms aecia and resting pycnidiospores on the lower surface (Figure 6.24). The aecia produce aeciospores that are wind-blown on to cereal plants, where the fungus penetrates the cereal indirectly via the stomata. Following this initial penetration of the cereal plant, the fungus produces a range of asexual spores known as urediniospores from a uredinium that has formed in the vicinity of the penetration site (Figure 6.25). The urediniospores are then disseminated by wind across distances of many hundreds of miles to infect new cereal plants. In favorable conditions the latent period is only about 7 days.

Treatment: The main control measure for cereal rusts is integration of the use of disease-resistant cultivars with a timely spraying program (see Chapter 10). Chemical treatment can be protective (Figure 5.7), using triazole fungicides, and/or curative, using succinate dehydrogenase inhibitor (SDHI) fungicides.

Figure 6.25 Light microscopy image of urediniospores of *Puccinia graminis*. **The arrows indicate the sites where the urediniospores are erupting through the epidermis.** (Courtesy of Joanne Knight, Plymouth University.)

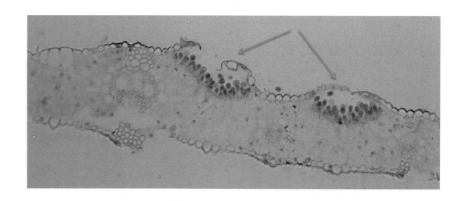

6.20 *Puccinia kuehnii*

Family: Pucciniaceae
Genus: *Puccinia*
Common name: Orange rust
Host: Sugarcane

This species of *Puccinia* is highly specific, and at the time of writing it is believed that it only infects sugarcane. Despite its host limitation, this pathogen is prolific. Early records suggest that its origins were in India, where it was characterized by E J Butler in 1914. However, widespread disease was first recorded in Australia in the 2000–2001 growing season. Orange rust is still causing major losses to the Australian agricultural economy annually, and is currently spreading rapidly across the tropics, including South-East Asia, the southern USA, Central and South America, the Caribbean Islands, and West Africa. Crop losses in fields infected with orange rust are estimated to be 20–40% of total yield.

Life Cycle: This fungus is believed to originate from wild *Saccharum* species that occur across a range of Pacific islands. The fungus initially spreads into the crop by wind-blown and water-splashed urediniospores, which are produced by the fungus when temperatures are in the range of 10–24°C. These spores settle on the leaves and leaf sheaths of the host plant (**Figure 6.26**) and subsequently germinate under conducive temperature and moisture regimes (15–25°C and 95–97% relative humidity). The total number of urediniospores produced is vast, with reports of one sorus producing 4700 spores, which are then wind-blown to start a new cycle of infection. Urediniospores occur on the lower leaf surfaces, are typically larger than those observed in brown rust pathogens, and are ovoid in shape. The fungus overwinters as teliospores, which are translucent, thin-walled, and do not have a **paraphysis** (a sterile hair-like filament). These spores then germinate in the spring, starting a new infection cycle.

Treatment: Control is generally achieved by using resistant cultivars, although in Australia this has been compromised by the evolution of a new strain of sugarcane orange rust fungus. However, there still remains a good resource of resistant cultivars for growers across the globe. There is the potential to produce a biological control agent using the parasitic fungus *Darluca filum*, but this has yet to be fully developed. In high-risk areas, chemical control using compounds such as tebuconazole and mancozeb is an option.

Figure 6.26 Orange rust (*Puccinia kuehnii*) on sugarcane. A color version of this figure can be found in the color plate section at the end of the book. (Courtesy of Josiane Takassaki, Ferrari – Laboratório de Doenças Fúngicasem Horticultura, Instituto Biológico.)

6.21 *Rhizoctonia solani*

Family: Ceratobasidiaceae
Genus: *Rhizoctonia*
Common name: Rice sheath blight, sugar beet root rot, potato black scurf (and a range of other depending on host specifics)
Main hosts: A broad range of host plants, mainly in the families Brassicaceae (formerly Cruciferae), Chenopodiaceae (mainly sugar beet), Solanaceae, and Poaceae (also called Gramineae)

Rice sheath blight is a fungal disease caused by *Rhizoctonia solani*. Infected stems develop lesions, the leaves dry out (**Figure 6.27**), and new tillers can also be destroyed. Subsequently, the disease can significantly reduce the leaf area and efficacy of the canopy, resulting in yield reduction. The disease is favored by dense stands with a heavily developed canopy, warm temperatures, and high humidity levels. In addition, excessive use of fertilizer tends to increase the damage caused by this disease.

Figure 6.27 (a) Rice plant stems showing sheath blight (*Rhizoctonia solani*) on main stems and sheath tissue, weakening the plants and reducing photosynthetic capacity, and (b) rice plant leaves infected with rice sheath blight. Loss of chlorophyll results in reduced starch accumulation in the grain. (Courtesy of Donald E Groth, LSU AgCenter Rice Research Station.)

(a)   (b)

Life Cycle: This fungus survives in the soil from one year to the next as a hard, weather-resistant structure called a sclerotium. Certain rice pathogens of *R. solani* have evolved the ability to produce sclerotia with a dense outer layer that allows them to float and survive in water. When in contact with a rice plant, the fungus grows out from the sclerotium and into the leaf sheath. Later, new sclerotia that have developed on infected stem surfaces fall from the plant to complete the life cycle. Sclerotia can remain viable in the soil for several years.

Treatment: Rice sheath blight is considered to be an economically important disease of rice, second only to rice blast. It is a cause of increasing concern for rice production, especially in intensified production systems. The disease can be controlled using medium-grain rice varieties, which are more resistant to sheath blight than most of the long-grain varieties. In addition, rationalizing fertilizer usage (adapted to the cropping season), together with the timely use of fungicide to treat seeds, is effective. Furthermore, a well-structured crop density, sown to a stand of 135–180 plants per square meter, along with careful control of weeds, will reduce yield losses to this disease.

6.22 *Pycnostysanus azaleae*

Family: Incertae sedis
Genus: *Pycnostysanus*
Common name: Rhododendron bud blight or bud blast
Main hosts: *Rhododendron ponticum* and numerous garden species, cultivars of the genus *Rhododendron*, and deciduous azaleas

This fungus produces brown to black buds (Figure 6.28) that lead to aborted flowers. The condition can be quite devastating to a bush, and is frequently a problem in southern England.

Life Cycle: The fungus is spread by a leafhopper (*Graphocephala fennahi*), which is a sap-feeding invertebrate that punctures the epidermis of host plants with its **stylet** and can transmit the sap-based infections to the plant. However, in the case of the rhododendron leafhopper, the bud blast infection is transmitted to the buds as they are developing. The female leafhopper

Figure 6.28 Rhododendron bud blast, showing the brown to black bud and the white pinhead coremia, which contain conidia. A color version of this figure can be found in the color plate section at the end of the book.

deposits her eggs into incisions made in the developing buds, and the fungus is then transmitted to the plant. Some evidence is now emerging that this method of disease transmission is inaccurate, and that the other method of transmitting the fungus to the plant would be through the stylet as the leafhopper feeds on the plant. On close examination of infected buds, white pinhead structures known as coremia can be observed. These are conidia-bearing fungal organs, and thus sources of inoculum. It is quite possible that the leafhopper picks up these conidia and then carries these inoculum sources to a neighboring rhododendron bush, where it infects new floral buds.

Treatment: The best approach is cultural control, which involves selecting resistant cultivars, and removing infected buds and burning them. During June and July, leafhopper numbers should be controlled either with chemical pesticides or, if the use of chemicals is unacceptable, by applying a soap-based spray to plants that have leafhoppers, as this will block the spiracles of the insects and thus kill them. Treatment of infected plants with liquid seaweed extract will help to strengthen their innate defense mechanisms, which may suppress the development of the disease.

6.23 *Rhynchosporium secalis*

Family: Incertae sedis
Genus: *Rhynchosporium*
Common name: Barley leaf blotch
Main hosts: Barley

The genus *Rhynchosporium* consists of five species that cause leaf blotch of numerous grass species, but the most economically important pathogen is *R. secalis*, which infects barley.

Life Cycle: The fungus survives on crop debris and is seed-borne. The disease is favored by wetter conditions, and is mainly visible from mid-canopy development onwards. Early spring symptoms are very rare. Symptoms are variable and look similar to those of *Septoria tritici*; however, the chlorotic patches of *R. secalis* have a necrotic border (Figure 6.29). Primary infections arise from rain-splashed pycnidiospores from infected crop debris, and then splash dispersal of conidia disperses the infection through the crop canopy. The disease cycle is rapid, and new cycles of infection can occur within 14 days of initial spore dispersal. High temperatures (25°C or higher) and dry conditions reduce the incidence of this disease.

Figure 6.29 *Rhynchosporium secalis.*
(a) Late-season symptoms on barley.
(b) The pycnidiospores. Note the decayed leaf tissue around the spores, which typifies this spore stage. (c) A conidium showing a short germ tube and attachment to the leaf epidermis.

(a)   (b)   (c)

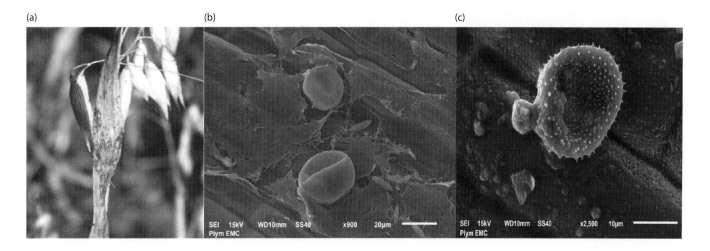

Figure 6.30 Distinct black tar-like lesions characteristic of acer tar spot (*Rhytisma acerinum*). A color version of this figure can be found in the color plate section at the end of the book. (Courtesy of Andrew Richards, formerly of Plymouth University.)

Treatment: An integrated approach is recommended, whereby growers rotate crops and bury crop debris with deep ploughing where possible. Incorporation of resistant cultivars into the cropping program is desirable, and timely applications of fungicides are crucial, with the key target being stem extension around growth stages 30 and 32. In high-risk areas where disease pressure is high in autumn-sown crops, low-dose fungicide applications may also be applied at earlier growth stages. The disease must not be ignored or overlooked, as yield losses can exceed 1.5 t/ha. Key fungicides include the strobilurins, chlorothalonil, and the succinate dehydrogenase inhibitor (SDHI) fungicides. Eradication is limited to the triazole compounds, such as prothioconazole.

6.24 *Rhytisma acerinum*

Family: Rhytismataceae
Genus: *Rhytisma*
Common name: Acer tar spot
Main host: Sycamore trees (specific races of the disease are found in other acer species)

Rhytisma acerinum is a highly specific fungal disease, very commonly seen on sycamore trees, and appearing from mid-summer until leaf fall as pronounced black blotches that resemble drops of tar—hence the common name "acer tar spot." These patches have a distinct yellow halo around them (Figure 6.30). The dark coloration is due to the fact that the spores are so densely packed together that no leaf tissue is exposed, apart from the stomatal guard cells (Figure 6.31).

Life Cycle: Shed leaves that have been infected during the previous season support apothecia that develop during late winter and early spring. These are saproxylic on the decaying leaves. Later in the spring, the fine ascospores that are released from the apothecia are carried back up to the newly developing leaves in air currents and breezes. These ascospores adhere to the leaves, causing infection. The first sign of infection is a pale yellow patch, which later develops into a raised **stroma**, and then blackens to form the characteristic black tar-spot lesion as the non-infectious conidiophores proliferate, with the initial yellowing remaining as a surrounding halo. These stromata only mature to form asci late in the season during leaf fall.

Treatment: There is no specific treatment for this disease, although infection can be limited by removal of dead leaves from around the base of the tree in winter. However, the disease is only weakly pathogenic and appears to cause no long-term damage to the plant. The tar spots prevent photosynthesis on the infected patches of leaf, and a heavy infection may cause early leaf fall, but ill effects are generally only minor.

6.25 *Zymoseptoria tritici* (syn. *Septoria tritici*)

Family: Mycosphaerellaceae
Genus: *Zymoseptoria*
Common name: Leaf blotch
Main hosts: Wheat

The correct taxonomic name for leaf blotch is *Zymoseptoria tritici*. This filamentous fungus belonging to the ascomycetes is a wheat plant pathogen that has co-evolved with the development of wheats during the last 10,000–12,000 years. The wheat-infecting lineage has probably emerged from a closely related wild type of *Mycosphaerella* pathogens that infect wild

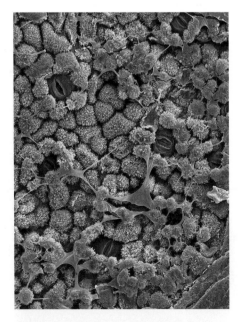

Figure 6.31 Acer tar spot (*Rhytisma acerinum*) magnified 1700×, showing very numerous, closely packed conidiophores. Note that no leaf tissue can be seen in the areas where the tar spots are present, except for the stomata that remain untouched by the spores. Note also the sticky secretion that may help to bind the conidiophores into a mass.

(a)   (b)

100μm ━━━

Figure 6.32 (a) *Septoria tritici* on the flag leaf of wheat (on the right), compared with a healthy flag leaf (on the left); (b) erupting pycnidia viewed using scanning electron microscopy. Note the tear in the epidermis of the leaf in the electron microscopy image.

grasses. Endemic progenitors of *Z. tritici* are still found on wild grasses in the Middle East, and these wild pathogens have a broader host range than the pathogen that infects wheat. Due to the widespread commercial acceptance of *Septoria* as *Septoria tritici*, this name will be adopted in this text.

Life Cycle: The pathogen survives the intercropping period as dormant mycelium, pycnidia, and pseudothecia on crop debris. Initial infections arise from wind-blown ascospores released from the pseudothecia, and from pycnidiospores released into rain splash. Both these spores are disseminated via rain splash, and germination of spores is optimal at temperatures of 15–20°C. The initial infection starts in the leaves of the lower canopy, and if left untreated the infection will travel vertically through the crop as pycnidia, but are also released from developing mycelium in the lower leaves. The released pycnidia are then rain splashed through the crop and eventually infect the flag leaf (Figure 6.32). The pathogen starts its life cycle in the host as a biotrophic pathogen, but once the long latent period (generally 15–18 days, but it can be as long as 28 days) is complete and the hyphae are established, the fungus switches to a necrotrophic life strategy.

Treatment: Yield losses can be very significant, and are in the range of 30–50% in high-risk areas such as the south-west of the UK. The strategy for managing *Septoria* is a combination of use of disease-resistant cultivars, crop rotations, and timely applications of fungicides (see Chapter 10) to manage the photosynthetic health of the main yield-forming leaves (leaf 3, leaf 2, and the flag leaf, which is known as leaf 1). The application of key fungicides involves use of a mix of chemical compounds with a range of site activity (known as a multi-site fungicide mix) to check the development of fungicide resistance in *S. tritici*.

6.26 Sooty Molds of Wheat

Family: Pleosporaceae and Davidiellaceae
Genus: *Alternaria* and *Cladosporium*, respectively
Common name: Sooty molds of wheat
Main hosts: Cereals

These pathogens infect all of the cereals. They grow on the surface of the **glumes** on the ears of ripening or ripened wheat (Figure 6.2b) during wet summers. The pathogens do not cause significant yield loss, but they can cause discoloration of the grain and surrounding tissue, and thus reduce the final quality characteristics of the yield and hence its marketability.

However, if the harvest is significantly delayed due to a persistent wet summer, the fungus can colonize the kernels, leading to the formation of small black lesions known as black point. Furthermore, if stored grain is not dried to 15–18% moisture content, the pathogen may cause post-harvest spoilage.

Life Cycle: Both *Alternaria* and *Cladosporium* produce hyphae that develop terminal conidia (Figure 6.2c), which are wind-blown or rain-splashed onto the ears post growth stage 75. The fungus then overwinters on crop debris.

Treatment: There is no chemical treatment. The main strategy is to ensure a timely harvest before summer conditions deteriorate into a wet autumn.

6.27 *Venturia inaequalis*

Family: Venturiaceae
Genus: *Venturia*
Common name: Apple scab
Main hosts: Apples, pears, and hawthorn

This is a key disease of apples, and it infects both the foliage (**Figure 6.33**) and the fruits. It can cause significant skin blemish on fruits (**Figure 6.34**), which leads to the fruit becoming downgraded, thus reducing marketing options for growers. In severe infections the fruits become so badly blemished that they have to be discarded.

Life Cycle: The pathogen overwinters as immature pseudothecia in plant debris left on the orchard floor. Pseudothecia complete their growth during the winter months, and as they mature they release ascospores, which are blown onto new leaf growth of the host as the spring develops. The ascospores also infect blossom, and this is a further source of fruit inoculum as these spores complete their growth into mycelium. The mycelium in living tissue is located just under the cuticle of the epidermis, where short, erect conidiophores produce distinctive one- or two-celled spiral-shaped conidia that can in turn spread the infection to fresh leaf and fruit tissue.

Treatment: Control consists of good cultural practice, including cleaning the orchard floor of windfalls and raking up the leaves following autumn leaf drop; the leaves should then be burned, not composted. Although it is time consuming, this practice will reduce the amount of overwintering inoculum and hence slow the course of any new-season infections. The use of disease-resistant cultivars is crucial, and modern scab-resistant cultivars of commercial varieties, such as the Braeburn apple, are being developed (**Figure 6.35**). Fungicides are available, but strict adherence to the withdrawal period is essential. However, for many small growers in the EU, the list of active chemical compounds is declining, so good cultural control and the use of resistant cultivars is likely to be the main strategy for controlling apple scab in small, non-commercial orchards.

Figure 6.33 Apple scab (*Venturia inaequalis*) on a leaf of apple (*Malus* species). Note the raised black patches which are composed of subcuticular mycelium and germinating conidia.

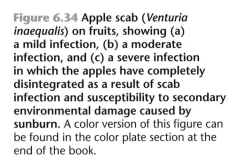

Figure 6.34 Apple scab (*Venturia inaequalis*) on fruits, showing (a) a mild infection, (b) a moderate infection, and (c) a severe infection in which the apples have completely disintegrated as a result of scab infection and susceptibility to secondary environmental damage caused by sunburn. A color version of this figure can be found in the color plate section at the end of the book.

(a)   (b)   (c)

Figure 6.35 The effect of apple scab (*Venturia inaequalis*) on a susceptible variety of Braeburn compared with a resistant variety. A heavy infestation of the disease in the susceptible variety is visible both (a) in the whole tree and (b) in a selection of the fruits from the same tree. The resistant variety of Braeburn shows a complete lack of infection in both (c) the whole tree and (d) the leaves and fruit from the same tree. A color version of this figure can be found in the color plate section at the end of the book.

Further Reading

Agriculture and Natural Resources, University of California. (2014) UC Pest Management Guidelines: Botrytis Fruit Rot on Strawberry. www.ipm.ucdavis.edu/PMG/r734100111.html

Agrios GN (2005) Plant Pathology, 5th ed. Elsevier Academic Press.

Amadi JE, Bamgbose O & Olahan GS (2012) The host range of *Aspergillus niger* and *Fusarium oxysporum* in the family Solanaceae. *Int J Agri Sci* 2:451–456.

Barakat FM, Adaba KA, Abou-Zeid NM & El-Gammal YHE (2014) Effect of volatile and non-volatile compounds of *Trichoderma* spp. on *Botrytis fabae* the causative agent of faba bean chocolate spot. *Am J Life Sci* 2:11–18.

Buczacki S & Harris K (2014) Collins Pests, Diseases & Disorders of Garden Plants, 4th ed. HarperCollins Publishers.

Byther RS (1987) Narcissus - Recommendations for Control of Basal Rot of Daffodils. Washington State University. http://horticulture. oregonstate.edu/system/files/onn110218.pdf

Forestry Commission (2015) Chalara dieback of ash (*Hymenoscyphus fraxineus*). www.forestry.gov.uk/ashdieback

Gabriel HB, Silva MF, Kimura EA et al. (2015) Squalestatin is an inhibitor of carotenoid biosynthesis in *Plasmodium falciparum*. *Antimicrob Agents Chemother* 59:3180–3188.

Kowalski T (2006) *Chalara fraxinea* sp. nov. associated with dieback of ash (*Fraxinus excelsior*) in Poland. *Forest Pathol* 36:264–270.

Kowalski T & Holdenrieder O (2009) Pathogenicity of *Chalara fraxinea*. *Forest Pathol* 39:1–7.

Linfield CA & Hanks GR (1994) A Review of the Control of Basal Rot and Other Diseases in Narcissus. Horticulture Research International. www.hdc.org.uk/sites/default/files/research_papers/BOF%2031%20Final%20Report.pdf

Louws FJ (2008) Major Strawberry Diseases. https://ncstrawberry.com/wp-content/uploads/2016/04/MajorStrawberryDiseases11-08.pdf

Mertely JC & Peres NA (2006) Botrytis Fruit Rot or Gray Mold of Strawberry. http://cloud.agroclimate.org/tools/deprecated/sas/publications/PP15200-BFR-revJul09.pdf

Royal Horticultural Society (2015) Narcissus Basal Rot. www.rhs.org.uk/advice/profile?PID=222

Selma MV, Freitas PM, Almela L et al. (2008) Ultraviolet-C and induced stilbenes control ochratoxigenic *Aspergillus* in grapes. *J Agric Food Chem* 56:9990–9996.

Sharma R (2012) Pathogenicity of *Aspergillus niger* in plants. *Cibtech J Microbiol* 1:47–51.

Tian J, Wang Y, Zeng H et al. (2015) Efficacy and possible mechanisms of perillaldehyde in control of *Aspergillus niger* causing grape decay. *Int J Food Microbiol* 202:27–34.

Visconti A, Perrone G, Cozzi G & Solfrizzo M (2008) Managing ochratoxin A risk in the grape-wine food chain. *Food Addit Contam Part A Chem Anal Control Expo Risk Assess* 25:193–202.

Weber RWS & Webster J (2002) Teaching techniques for mycology: 18. *Rhytisma acerinum*, cause of tar-spot disease of sycamore leaves. *Mycologist* 16:120–123.

Zhang J, Wu M-D, Li G-Q et al. (2010) *Botrytis fabiopsis*, a new species causing chocolate spot of broad bean in central China. *Mycologia* 102:1114–1126.

Chapter 7
Bacteria Diseases

Of the millions of bacteria that exist on our planet, surprisingly few species are associated with plants, and of those that are, most are non-pathogenic, and they are often mutualistic or saprophytic. It is not known how many bacteria are actually pathogenic to plants, and any estimates would become out of date almost immediately as these organisms are constantly baffling pathologists due to the rate at which they can adapt to changing conditions and ultimately evolve into new species. However, a number of genera, such as *Xanthomonas*, *Pseudomonas*, and *Erwinia*, do have representatives within their ranks that provide an ever present threat of disease to plants.

Although they are relatively few in number compared with fungal pathogens, bacterial diseases of plants are ubiquitous and can devastate important crops, including staples such as potatoes, rice, and cassava, among many others. Plant pathologists, geneticists, and plant breeders are fighting an ongoing battle to find new varieties of resistant crop plants, soil scientists are working on ways to improve soil quality and condition, and technologists are looking at new ways to lift, transport, pack, and store crops to avoid damage that may lead to infection.

In this chapter we highlight a selection of representative bacterial diseases. Students should bear in mind that these are representatives only, and that a different strain of the same species may have different effects, induce different symptoms, and vary in its virulence.

7.1 *Agrobacterium tumefaciens*

Family: Rhizobiaceae
Genus: *Agrobacterium*
Classification: Gram-negative rod
Common name: Crown gall disease
Main hosts: Trees, including stone fruit trees, and a wide range of herbaceous eudicot plants

Many members of the family Rhizobiaceae are commonly associated with nitrogen fixation in the root nodules of legumes—a mutualistic symbiosis. However, species belonging to the genus *Agrobacterium* are pathogenic to plants. *A. tumefaciens* is the most well-known and well-studied species in this group. As the name implies, this pathogen induces tumor-like growths in the host that can develop to a significant size as galls, especially on the trunks of trees such as the oak (*Quercus* species) (Figure 7.1).

7.1 Exploitation of Ti Plasmids
The extraordinary capacity of *Agrobacterium tumefaciens* with its Ti plasmids to infect and integrate with plant host DNA has been extensively exploited by plant geneticists. Indeed the use of sections of the plasmid DNA that exclude the tumor-forming sections of the strand has been shown to be highly precise and reproducible, paving the way for plant genetic research.

Figure 7.1 Crown gall tumor caused by *Agrobacterium tumefaciens* **on oak tree** (*Quercus robur*).

Life Cycle: Like the non-pathogenic members of this family, *A. tumefaciens* is a soil bacterium that is attracted to plant root exudates that it detects in the rhizosphere. These bacteria have flagella, and actively swim towards the attractants by the process of chemotaxis. Pathogenicity is dependent on the presence of tumor-inducing plasmids (Ti plasmids) that are found within the cells. Plasmids are extra-chromosomal entities that exist within many cell types. In *A. tumefaciens* we see proactive plasmids that give their hosts a dual advantage—they represent the pathogenic characteristic of the bacterial host, and they also facilitate conjugation, and thus horizontal gene transfer, within the bacterial community, thereby expanding the capacity for pathogenicity.

Initial entry into the plant host is via a wound, and then a section of plasmid tDNA is transferred to the host plant cell via specially formed pili. This then integrates with the plant host DNA, inducing cell proliferation (leading to the formation of gall tumors), thereby increasing both the capacity and the ingress of nutrients to the infected tissue. The galls are induced to produce **opine** compounds that can then be accessed by the bacterium as a nitrogen source. Although each plasmid has its own complement of circular DNA, and replicates independently of the bacterial host cells, replication of bacterium and plasmid tends to be synchronous as colonies develop throughout the tumorous tissues.

Treatment: This is a difficult disease to control, although the use of *A. tumefaciens* from which the plasmids have been removed has shown some success, as this offers a degree of competition with the infectious bacteria. Copper treatments can also be used, although these can cause damage to the plant as well as the pathogen.

7.2 Bacillus pumilus

Family: Bacilliaceae
Genus: *Bacillus*
Classification: Gram-positive rod, obligate aerobe
Common name: Mango twig dieback and leaf blight
Main host: Mango (*Mangifera indica*), other stone fruits, cucurbits, onions

The mango is a popular commercial fruit that is grown in the tropics and subtropics. All parts of the mango plant can be affected by a range of fungal, bacterial, viral, and algal diseases, and the growing global popularity of the fruit means that there is a strong impetus to combat infectious diseases in this plant. *B. pumilus* has only been associated with disease in the mango in recent years, and is not yet considered to be a serious threat. However, the bacterium is sometimes used as an antagonist against a number of fungal pathogens, including *Fusarium* species. Thus if plants are found to be infected with the bacterium in the form of a disease, this reduces the potential for the bacterium to be used as a biocontrol agent.

Life Cycle: Recent studies conducted in Egypt (Galal et al. 2006) indicate that infection occurs specifically following damage to or wounding of plant tissues. Inoculants in the form of foliar sprays or additives into soils did not induce disease. In addition, different mango varieties showed varying levels of susceptibility.

Treatment: Cultural control consists of selection of resistant varieties of mango tree, careful handling to avoid causing damage that would provide an entry point for infection, and the implementation of strict hygiene measures.

Further studies of the effects of *B. pumilus* are needed (and possibly the development of attenuated strains) before it is used as a treatment against fungal pathogens.

Note: Biologists in Ukraine have recently identified *B. pumilus* as a causal agent of disease in Scots pine trees (*Pinus sylvestris*). The pathogen causes soft rot symptoms in seedlings, wetwood symptoms in young trees, and bleeding lesions in older trees.

7.3 Burkholderia caryophylli

Family: Burkholderiaceae
Genus: *Burkholderia*
Classification: Gram-negative rod
Common name: Carnation bacterial wilt
Main host: Carnations (*Dianthus caryophyllus*)

The Burkholderiaceae were once classified as part of the family Pseudomonadaceae, but modern techniques have separated the members of this group into their own family. *Burkholderia* species tend to be fairly specific to a small range of plant hosts, although oddly a number of species in this group, known as the *Burkholderia cepacia* complex, can cause quite serious diseases in humans and are often associated with hospital-acquired infections (this capacity for cross-kingdom infection is most unusual). Carnation wilt causes wilt, foot rot, and root rot symptoms in the genus *Dianthus* and a few other horticulturally important groups, including gypsophila and statice.

Life Cycle: This bacterium is classified as soilborne and can enter the host via the roots. However, spread of the disease is more often associated with contamination via cuttings, which are the preferred method of industrial-scale propagation. The bacterium can remain dormant for long periods, so that by the time symptoms are detected the infection is fully established.

Treatment: Carnations are used extensively in the horticultural industry, and this disease has been combated by selective breeding techniques and good hygiene practices in the EU, where no cases have been reported for the last 25 years. However, carnation wilt continues to pose a significant threat to the Asian horticultural industry, where it is still regarded as an important soilborne disease, and Japanese plant pathologists are working on isolation methods to combat its effects. Furthermore, some of the original species plants that are growing wild in Mediterranean regions are still susceptible to the disease.

7.4 Clavibacter michiganensis subsp. michiganensis

Family: Microbacteriaceae
Genus: *Clavibacter*
Classification: Gram-positive rod, aerobe
Common name: Tomato vascular wilt, potato ring rot
Main hosts: *Solanum* species, especially tomatoes, potatoes, aubergines, and peppers

C. michiganensis is a globally widespread bacterium that causes a range of disease symptoms in *Solanum* species. There are only a very few species belonging to this genus (*C. xyli* infects sugarcane), of which *C. michiganensis* subsp. *michiganensis* is the most well studied, largely due to the fact that it infects commercially important crops such as tomatoes and potatoes. In both of these hosts the disease causes lesions, dieback, whole-plant dwarfing, canker on older stem tissue of tomatoes, and ring rot in potato tubers. In extreme infections, necrosis results in significant crop losses.

Life Cycle: The bacterium is typically seed-borne, and therefore even though it does not sporulate, it easily survives from one host generation to the next. Infection can also be induced during transplantation. Once the plant is infected the disease travels through the vascular tissue. The genes associated with disease are plasmid encoded; without the plasmids, the bacterium becomes an endophyte, with complete loss of its pathogenic virulence.

Treatment: Cultural control consists of good hygiene practices during cultivation, and prompt removal of debris, especially debris that contains seed material.

7.5 Erwinia amylovora

Family: Enterobacteriaceae
Genus: *Erwinia*
Classification: Gram-negative rod, aerobic (facultative anaerobe)
Common name: Fire blight
Main hosts: Trees and shrubs, mainly from the family Rosaceae

Members of the family Enterobacteriaceae are more commonly associated with animal infections. However, the *Erwinia* species typically infect plants. There are a number of species which affect a range of hosts, but *E. amylovora* is by far the most common, widespread, and commercially significant, causing blight and wilt symptoms in all parts of the plant. Pear trees, apple trees, and some berry fruits are the main commercially important crops affected, together with some ornamentals, such as *Pyracantha*. The common name "fire blight" alludes to the appearance of affected trees, which appear at first to be on fire (Figure 7.2), and then later appear blackened as if scorched. Affected tissues may also exude a pale-colored ooze, especially in the morning when the host water potential is high. Fire blight originated in North America, and is now widely distributed across most of the northern hemisphere, especially Europe; it has also gained a foothold in New Zealand.

Life Cycle: Natural infection occurs mainly via the flowers, and those first infected appear water soaked. The bacterium may also enter the host through wounds. If left unchecked, the bacterium replicates rapidly and spreads through the vascular tissue to the whole plant. Infection may also occur during cultivation, particularly during grafting; if certified clean stock is not used the bacterium can travel rapidly from the scion into the rootstock tissues. The bacterium overwinters in a dormant state until the new growth season begins.

Treatment: This is a difficult disease to tackle. Prophylactic antibiotic treatment was formerly used, but is now largely banned. The bacterium often developed resistance to such treatments over time, rendering them ineffective. Recent research indicates that there may be a viable alternative method of introducing antibiotics, along with resistance inducers, directly into the trunk of the host plant, possibly overwhelming the disease while the systemic acquired resistance (SAR) response is initiated. Indeed compounds that form part of the SAR response have been found to rapidly accumulate in the canopy (Aćimović et al., 2015). Removal of the first infected flowers may be effective if it is done promptly, although this approach may not be practicable in large-scale plantations. Early removal of infected plants, together with the use of certified clean stock, is an effective form of cultural

Figure 7.2 *Erwinia amylovora* **(fire blight) is a common bacterial disease of trees and shrubs, especially species belonging to the family Rosaceae, such as apples (Malus species shown here), pears, hawthorn, and many others.** A color version of this figure can be found in the color plate section at the end of the book. (Courtesy of Robert L Anderson, formerly of USDA Forest Service, Bugwood.org)

control. Use of bacterial antagonists such as *Bacillus subtilis* and a number of non-pathogenic *Pseudomonas* species has also been demonstrated to be effective. These antagonists not only crowd out *E. amylovora*, but may also produce antibiotic substances that outcompete the pathogen.

7.6 *Pectobacterium atrosepticum*

Family: Enterobacteriaceae
Genus: *Pectobacterium*
Classification: Gram-negative rod, aerobic (facultative anaerobe)
Common name: Potato blackleg, potato soft rot
Main host: Potatoes

Although *Pectobacterium atrosepticum* is largely associated with diseases in potatoes, it has a wide host range, including many commercially important crops such as brassicas, bananas, soya, and many horticulturally important plants, and the disease has a global distribution. The symptoms are clearly visible as the disease causes soft rot, wilting, lesions, necrosis, and stunted growth. Potato crop losses due to this disease can be significant.

Life Cycle: The disease typically infects the tuber first, causing vascular discoloration, rotting in the area around the stolon, and lesions associated with the **lenticels**. In extreme cases the pathogen may spread to the rest of the tuber, and ultimately to the stems, causing blackening—hence the common name "blackleg." Infection may also develop from wounded aerial parts of the plant via insect vectors.

Treatment: This is a difficult disease to treat. Lenticel cell walls can become suberized (hardened and made corky by the deposition of suberin, a natural waterproofing process in higher plants), making treatment with water-based chemicals ineffective. In the past, sulfuric acid was used, but this was associated with health risks both to the plant and to the end consumer. The most effective form of control is prevention, by use of clean stock materials, planting more resistant potato varieties, adopting strict hygiene measures, and removal of plant debris. Early removal of diseased plants may prevent spread of the pathogen.

Note: Recent studies indicate that this pathogen can interact with endophytic bacteria present in the host, often resulting in increased pathogenicity. Furthermore, these endophytes, although normally commensal or mutualistic with the host (for example N_2-fixing bacteria), may become opportunists during pathogenic infection by exploiting substances (for example, sugars) that are released during rotting. Indeed, pathogenicity appears to continue unchecked even when the endophytic bacteria predominate in the microbial community within the host.

7.7 *Pseudomonas syringae*

Family: Pseudomonadaceae
Genus: *Pseudomonas*
Classification: Gram-negative rod
Common name: Bacterial canker
Main hosts: A wide range of host plants, but prevalent on trees, especially fruit trees

Canker diseases are typically caused by fungi, although the few that are caused by bacteria are economically significant. Furthermore, whereas cankers caused by fungi are normally only associated with their effects on the

7.2 Cloud Seeding

The ice-nucleation function of *Pseudomonas syringae* is believed to have natural cloud-seeding properties that have an impact on global weather patterns. Recent studies have implicated conflicting historical roles for this bacterium—as a potentially devastating plant pathogen on the one hand, and as a cloud-seeding organism that helps to bring life-giving rain to agricultural systems on the other.

Figure 7.3 Apple fruit with the small distinctive lesions caused by *Pseudomonas syringae*.

woody tissues of plants (causing large scabby lesions on trunks and twigs, which often exude a gum), the bacterial canker pathogens also affect all of the other aerial parts of the plant. Buds and flowers exhibit blight symptoms that are almost indistinguishable from the effects of *Erwinia* species, whereas the leaves and fruits have distinct small, often sunken, lesions over the surface (Figure 7.3). Dieback symptoms may also result from infection.

Life Cycle: *Pseudomonas syringae* (named after the lilac tree, belonging to the genus *Syringa*, in which it was first identified) can be transmitted by insects, wind, water splash, and infected tools, or may persist during cultivation. Entry to a new host is generally through wounds or via natural openings such as stomata or lenticels. Most pathovars produce a cytotoxin called **syringomycin** that kills areas of host tissue, where the bacterium can then multiply by binary fission. *P. syringae* also has **ice-nucleation** properties that cause cellular damage, providing further areas for bacterial multiplication. This tendency to multiply in areas of dead host tissue relates to the bacterium's ability to switch to saproxylic nutrition when conditions are not favorable for pathogenesis. *P. syringae* produces a suite of host-resistant virulence effector proteins. It is also able to detect competitors via **quorum sensing**, and can initiate production of a **biofilm** to outcompete these organisms.

Treatment: Copper sprays are effective for some infections, although these sprays can also damage the host tissue and their use is undesirable in commercial crops. In addition, some pathovars are resistant to copper. Overall, this is a very difficult pathogen to treat, and it is combated mainly by management. Good sanitation together with the use of clean stock reduces the level of infection. Pruning when temperatures are above freezing, to avoid ice nucleation before the wounds have healed, is also good practice.

7.8 *Ralstonia solanacearum*

Family: Burkholderiaceae
Genus: *Ralstonia*
Classification: Gram-negative rod, flagellated aerobe
Common name: Bacterial wilt (this name is also applied to other bacterial diseases), Granville wilt in tobacco
Main hosts: A wide range of host plants, especially members of the family Solanaceae, and bananas

Ralstonia species were originally classified as belonging to the genus *Pseudomonas*, but have since been reclassified due to pigmentation differences. *R. solanacearum* is one of the most significant bacterial pathogens of plants worldwide. Although its distribution is strain and race dependent, it is typically more common in tropical and subtropical regions It has a wide host range that includes many major food crops, such as potatoes, tomatoes, peppers, aubergines, ginger, tobacco, and bananas, among many others. As such, it is one of the most well-studied bacterial plant pathogens, and it has been shown to have a particularly large number of effector proteins associated with its pathogenic virulence.

Note: *R. solanacearum* infection in bananas is known as Moko disease, and is responsible for serious economic losses in tropical regions. The specific race (race 2) is thought to have evolved from wild *Heliconia* plants. The disease is transmissible via soil, seedlings, infected plant tissues, insect vectors, cultivation tools, and transportation of crops. The disease affects all parts of the plant, causing wilt symptoms in leaves (following collapse of the petiole), flowers, and roots. It also causes early ripening, stunting, and rotting of fruits from within (Figure 7.4). There is no known treatment, so all infected plants

Figure 7.4 Banana fruits infected with Moko disease (*Ralstonia solanacearum*). Note the stunted development, and the dark coloration indicating rotting from within the fruits. (Courtesy of Luadir Gasparotto, Creative Commons Attribution 3.0 Australia License.)

must be destroyed. Good sanitation and swift removal of infected plants, together with a 5-m *cordon sanitaire*, are the only effective control measures.

Life Cycle: This is a highly opportunistic motile soil bacterium. It is attracted to host plant root exudates, and then enters the host by any route available where damage has occurred to plant tissue, often favoring a direct route into the xylem following attack by sap-sucking bugs. In the xylem it multiplies by binary fission until the numbers of bacteria are so great that the vascular system becomes compromised and the host tissue exhibits characteristic wilting symptoms. Virulence and host resistance response factors are highly coordinated in this species. Such autoinduction, which is sometimes called quorum sensing, is a highly complex system of gene expression-inducing signaling molecules.

Treatment: There are few effective control methods available other than preventive measures. These include good sanitation, use of clean stock plants (certified disease-free), crop rotation systems that alternate host plants with non-host plants, and the use of plants that have been bred for resistance (although to date no host plants are considered fully resistant). The use of natural tannins to inhibit quorum sensing is currently being investigated.

7.9 *Streptomyces scabies*

Family: Streptomycetaceae

Genus: *Streptomyces*

Classification: Gram-positive filamentous, aerobic

Common name: Potato common scab

Main hosts: Potatoes and other tuberous and root vegetables, such as beets and carrots

Potatoes are one of the most important staple foods globally, and a huge amount of research is focused on the maintenance of healthy potato crops. Although a number of pathogens cause scab on potato tubers, *Streptomyces scabies* is one of the most significant species. The disease does not kill the plant or prevent the development of tubers, and indeed there is only minimal yield reduction as a result of infection. However, the scabby lesions that are caused by the pathogen do affect the marketability of the crop, reducing the number of Grade 1 yields. The most nutritious part of the tuber is just beneath the skin, and this part will be discarded when peeled to remove the damaged tissue.

Life Cycle: The name *Streptomyces* literally means "twisted fungus", and this bacterium was once classified as a fungus because of its filamentous structure. Like a fungus it forms hyphal mycelia, although these are typically much smaller than fungal mycelia. It is a soilborne pathogen so it most commonly affects root and tuberous crops, entering the plant through the lenticels or via damaged tissue.

Treatment: The most effective control measures are strict hygiene protocols when handling plants and tools, and the use of crop rotation methods that include non-root crops.

7.10 *Xanthomonas campestris*

Family: Xanthomonadaceae

Genus: *Xanthomonas*

Classification: Gram-negative rod, obligate aerobe

Common name: Black rot

Main hosts: A wide range of host plants, including members of the family Brassicaceae

7.3 Secondary Metabolites from *Streptomyces* Species: A Range of Functionality

Streptomyces is a vast genus containing over 500 known species. Most of these are saproxylic, although a few species, including *S. scabies*, are pathogenic. *Streptomyces* species typically form secondary metabolites. These are complex biochemical compounds, usually in the form of antibiotics, designed to enable the bacterium to outcompete other soil microorganisms—and utilized by humans for medical purposes. Pathogenic species also produce secondary metabolites, in this case toxins known as thaxtomins, which are responsible for the virulence of the pathogen. The genes associated with the synthesis of these toxins are uniquely clustered on a segment of chromosome known as a "pathogenicity island."

7.4 Common Pathogenic Bacterium Used in Everyday Foods

Xanthomonas campestris is the source of the polysaccharide xanthan gum that is utilized extensively as a thickening agent or as a stabilizing agent in foods and other commonly used products.

Figure 7.5 Cabbage leaf showing infection by *Xanthomonas campestris* pv. *campestris*.

The most common diagnostic symptom of *Xanthomonas campestris* infection is spotting on the leaves of infected host plants. In extreme cases, the disease advances through the vascular tissue, causing severe damage to other parts of the plant, which may ultimately result in death. Hosts are pathovar specific. For example, *X. campestris* pv. *campestris* (Xcc) is a common disease of brassicas (Figure 7.5), and *X. campestris* pv. *phaseoli* affects beans. However, *X. campestris* pv. *graminis* specifically infects grasses such as the bents, causing bacterial wilts in these hosts rather than the more typical leaf spot symptoms.

Xcc affects a whole range of brassica crops, and causes significant economic losses globally. Other pathovars are also wide ranging, limited only by their specific host's habitat requirements.

Life Cycle: Like most bacteria, *X. campestris* replicates by binary fission. It is favored by warm, humid conditions, when rapid replication ensures its success. It can survive in plant debris after harvest, and may remain dormant in host seed until germination, when it travels through the vascular system to the leaves and often also to other tissues, causing damage to the xylem as it does so. In the leaves of many brassicas, water is exuded by **guttation**, carrying the pathogen with it. This allows the pathogen to spread rapidly through crops by wind, by water splashing during rain or irrigation, or via insect vectors.

Treatment: This ubiquitous, virulent pathogen is difficult to treat. The use of certified clean seed stock and implementation of crop rotation practices decrease infection rates. However, many weed species are also hosts for this disease, allowing it to maintain a presence despite the best efforts of the grower. Extensive genetic studies of both the bacterium and a range of host organisms (including the weedy brassica *Arabidopsis thaliana*, which is used extensively as a model plant in genetic studies, and was the first plant to be fully gene sequenced) have been conducted. The aim of this research is to identify genes associated with secretory systems, pathogenicity, and virulence in the pathogen, and with resistance mechanisms in the host plants, with the long-term goal of manipulating crop resistance.

7.11 *Xanthomonas fragariae*

Family: Xanthomonadaceae
Genus: *Xanthomonas*
Classification: Gram-negative rod, obligate aerobe
Common name: Strawberry angular leaf spot
Main host: Strawberry plants (*Fragaria* × *ananassa*)

Strawberries are an important cash crop globally, but are extremely susceptible to many fungal and bacterial diseases. *Xanthomonas fragariae* is a highly infectious bacterium that affects both wild and cultivated strawberries. It has a widespread distribution, being found across the Americas, parts of Africa, Asia, Australasia, and parts of Europe, although it is currently classified as eradicated in the British Isles. The bacterium causes spotting and lesions of the upper surface of the leaves (Figure 7.6), and a viscous exudate from the underside of the leaves that dries to a white scaly film. It also affects the crown, stems, sepals, and fruits.

Life Cycle: *X. fragariae* overwinters in a dormant state (known as **sporulation**) in dead strawberry plant debris, and can survive in this state for long periods while awaiting favorable environmental conditions. It typically

Figure 7.6 Strawberry leaf showing characteristic spotting symptoms of *Xanthomonas fragariae* infection. (Courtesy of U Mazzuchi, University of Bologna, Bugwood.org)

infects new hosts via water splash, especially from overhead irrigation systems. Once the pathogen has infected a plant, it travels via the vascular system to affect all of the aerial parts of the plant. The disease can also be transmitted during cultivation, and proliferates in warm humid conditions.

Treatment: Control measures include sourcing disease-free stock plants from registered suppliers, maintaining strict hygiene when handling plants, providing adequate ventilation to lower the relative humidity, avoiding the use of overhead irrigation systems, and prompt removal of plant debris. Copper-based treatments can be used in extreme cases, although repeated doses can damage the host tissue, so it is preferable to avoid their use in commercial crops.

Note: A similar disease, known as cucurbit angular leaf spot, is caused by *Pseudomonas syringae* (see above), a broad-spectrum bacterium with numerous hosts and a range of disease effects.

7.12 *Xylella fastidiosa*

Family: Xanthomonadaceae
Genus: *Xylella*
Classification: Gram-negative rod, obligate aerobe
Common names: Olive tree decline, phoney peach disease, Pierce's disease (in grapevines), citrus variegated chlorosis
Main hosts: Olives, drupes (for example, peach), grapes, citrus

X. fastidiosa causes a devastating bacterial disease of many plants, especially commercially grown fruit trees, but also a number of economically important horticultural shrubs, such as coffee, rosemary, and hebe. Olive groves in Apulia, Italy have been the worst hit in Europe, and recent cases have been reported in the French island of Corsica and in mainland France. The European Union (EU) has dubbed the European strain of this disease (CoDiRO) the most dangerous plant bacterium in Europe. In southern states of the USA, Central America, and northern parts of South America, olive trees are unaffected. This is because a different strain of the disease, Multiplex, is found in these regions, and this strain has a completely different host range (for example, peach). In the UK, this disease primarily infects broad-leaved trees, although as yet the disease has not established itself as a major threat. However, this situation may change!

Life Cycle: *X. fastidiosa* is transmitted by insect vectors, specifically species belonging to the order Hemiptera (the true bugs), such as aphids, leafhoppers, and cicadas. The insects feed on plant sap, simultaneously transmitting the pathogen, which then travels through the host's xylem tissue. As the bacterium replicates, the xylem vessels become blocked both by bacterial aggregates and by gums produced by the host's own defense mechanisms. Eventually the plant's fluid transport systems cease to function, resulting in the death of the plant.

Treatment: As yet there is no effective treatment. Complete removal of infected plants, followed by disinfection, is currently the only course of action. Much research is currently addressing the question of how to tackle this disease.

Further Reading

Aćimović SG, Zeng Q, McGhee GC et al. (2015) Control of fire blight (*Erwinia amylovora*) on apple trees with trunk-injected plant resistance inducers and antibiotics and assessment of induction of pathogenesis-related protein genes. *Front Plant Sci* 6:1–10.

Agrios GN (2005) Plant Pathology, 5th ed. Elsevier Academic Press.

Eichenlaub R & Gartemann K-H (2011) The *Clavibacter michiganensis* subspecies: molecular investigation of Gram-positive bacterial plant pathogens. *Annu Rev Phytopathol* 49:445–464.

European Food and Safety Authority, Panel on Plant Health (PLH) (2013) Scientific Opinion on the risk to plant health posed by *Burkholderia caryophilli* for the EU territory with the identification and evaluation of risk reduction options. *EFSA J* 11:3071.

Galal AA, El-Bana AA & Janse J (2006) *Bacillus pumilus*, a new pathogen on mango plants. *Egypt J Phytopathol* 34:17–29.

Henke JM & Bassler BL (2004) Bacterial social engagements. *Trends Cell Biol* 14:648–656.

Jones DG (1987) Plant Pathology: Principles and Practice. Open University Press.

Kawanishi T, Uematsu S, Nishimura K et al. (2009) A new selective medium for *Burkholderia caryophilli*, the causal agent of carnation bacterial wilt. *Plant Pathol* 58:237–242.

Köiv V, Roosaare M, Vedler E et al. (2015) Microbial population dynamics in response to *Pectobacterium atrosepticum* infection in potato tubers. *Sci Rep* 5:11606.

Kovaleva VA, Sholovylo YI, Gorovik YN et al. (2015) *Bacillus pumilus* – a new phytopathogen of Scots pine – Short Communication. *J Forest Sci* 61:131–137.

Lerat S, Simao-Beaunoir A-M & Beaulieu C (2009) Genetic and physiological determinants of *Streptomyces scabies* pathogenicity. *Mol Plant Pathol* 10:579–585.

Morris CE, Monteil CL & Berge O (2013) The life history of *Pseudomonas syringae*: linking agriculture to earth system processes. *Annu Rev Phytopathol* 51:85–104.

Vicente JG & Holub EB (2013) *Xanthomonas campestris* pv. *campestris* (cause of black rot of crucifers) in the genome era is still a worldwide threat to brassica crops. *Mol Plant Pathol* 14:2–18.

Willey JM, Sherwood LM & Woolverton CJ (2017) Prescott's Microbiology, 10th ed. McGraw-Hill Education.

Zambryski P (2013) Fundamental discoveries and simple recombination between circular plasmid DNAs led to widespread use of *Agrobacterium tumefaciens* as a generalized vector for plant genetic engineering. *Int J Dev Biol* 57:449–452.

Useful Websites

CABI. Invasive Species Compendium. www.cabi.org/isc/datasheet/21908

CABI. Plantwise. www.plantwise.org

Cornell University. Cornell Fruit Resources: Resources for Commercial Growers. www.fruit.cornell.edu

Olive Oil Times. www.oliveoilnews.com

The Banana Board. www.thebananaboard.org

Chapter 8
Viral Diseases

Although plant viruses rarely kill their hosts, they can cause substantial damage, not only to crop yields but also to plant morphology—in the horticultural industry unsightly plants do not attract buyers! In this chapter we discuss a number of different plant viruses, specifically focusing on those of commercial importance.

The potyviruses are one of the most important groups of plant viruses. Potyviruses (ssRNA+) are filamentous obligate plant viruses that form long (up to 900 nm), narrow (only up to 15 nm in diameter), non-enveloped structures with helical symmetry. They are the largest taxonomic group of plant viruses identified to date, and represent some of the most economically important viral pathogens of plants today. Around 25% of all plant virus species belong to this group. The name originates from potato Y virus (also known as potato virus Y), after the host in which the virus was first recorded. However, we now know that the host range for these viruses is vast, exceeding 2000 plant species, including numerous agriculturally and horticulturally important plants.

Not only do the potyviruses include large numbers of virus species and large numbers of host species, but also the morphological effects of potyvirus infection on the hosts are highly diverse (for example, mottling, streaking, lesions, bunching, thickening, stunting, and color-breaking). Like many viruses, potyviruses induce the development of inclusion bodies within the host, and these are often large enough to be seen using a light microscope. Although potyvirus-induced inclusion bodies are also variable, and include amorphous, crystalline, and cylindrical forms, it is the pinwheel form that is a characteristic feature of this group (Figure 8.1).

The vast majority of plant viruses are single-stranded RNA (ssRNA), and most non-ssRNA virus families tend to be smaller groups with limited host ranges. However, the geminivirus family is unusual in that it is a large and commercially important group of ssDNA viruses with many species, a wide host range (including many tropical and subtropical staples, such as cassava and maize), and diverse host symptoms. Many species in this group induce a typical "curly-top" effect, although streaks and mosaics are also common. The geminiviruses are a distinctive group consisting of paired capsids each with dimensions of around 20 × 30 nm. One of the reasons for the success of this family is its high rate of viable spontaneous recombination. This gives the viruses a tactical advantage in the arms race with their hosts, which often fail to keep up with their constantly evolving attackers. As a result, geminiviruses can spread rapidly through a crop.

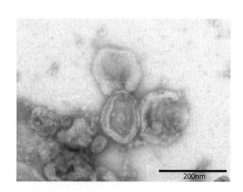

200nm

Figure 8.1 Pinwheel inclusion body found in a lettuce plant infected by a potyvirus. (Courtesy of Jason Mulvinney, formerly of Plymouth University.)

The two families described above represent a high proportion of the viruses that affect our agricultural and horticultural plants, although the list is hardly exhaustive. This chapter highlights some of the more common species from these and a number of other plant virus families.

8.1 African Cassava Mosaic Virus (ACMV)

Family: Geminiviridae
Genus: *Begomovirus*
Main hosts: Members of the family Euphorbiaceae
Structure: ssDNA, circular
Transmission: Via whitefly

The root of the cassava plant (*Manihot esculenta*) is one of the major carbohydrate sources in tropical regions and some subtropical regions globally, especially in Africa, Asia, and South America (from where it originates). Around 11% of the world's population use cassava as part of their staple diet. The plant's use in ethanol production is also gaining prominence in many developing countries, and thus its economic importance cannot be overstated. Cassava is affected by a number of diseases, the most important of which is African cassava mosaic virus (ACMV) (Figure 8.2). More than 50% of the world's cassava is produced in African states, and therefore ACMV poses a significant threat to food security.

ACMV is a geminivirus (discussed earlier in the chapter) and has a fairly limited host range, affecting mainly members of the family Euphorbiaceae (to which cassava belongs). The virus specifically infects the leaves, causing the mosaic effect, significant yellowing, and a reduction in leaf size. As a result, photosynthesis is substantially reduced, and the decreased amount of photosynthates available for starch production leads to a reduction in development of the storage roots.

Treatment: Roguing can be used where the disease incidence is low. As ACMV is transmitted by whitefly, planting when whiteflies are less abundant reduces the incidence of disease. However, significant disease outbreaks have also resulted from vegetative propagation, so newer techniques using more rigorous sanitation measures when taking cuttings, as well as the use of tissue culture methods, are currently being promoted by agriculturalists to tackle this disease. Resistant cultivars are also currently being developed.

For a description of cassava brown streak disease, see Research Box 4.1 in Chapter 4.

(a) (b)

Figure 8.2 Cassava plant, the roots of which are a major carbohydrate source throughout the tropics. (a) Leaves of healthy plants; (b) leaves affected by cassava mosaic virus—note the symptoms of this disease, namely misshapen leaves that are mottled with the distinct mosaic pattern. (Courtesy of Biosciences for Farming in Africa.)

8.2 Barley Yellow Dwarf Virus (BYDV)

Family: Luteoviridae

Genus: *Luteovirus*

Main hosts: Cereals and other grasses

Structure: ssRNA+, non-enveloped, isometric

Transmission: Via insects (mainly aphids)

BYDV is possibly the most economically important of all plant viruses. This pathogen is named for its effects on barley, but can also seriously damage most other cereal crops, such as wheat, oats, rice, and maize, causing yield losses of up to 50%, depending on the virus strain and host strains.

Infection in a crop appears as often large, patchy areas where the plants are smaller and less vigorous and are distinctly yellowing due to reduced chlorophyll production. The difference between a healthy plant and an infected one can be clearly seen in Figure 8.3, which shows plants that are still at an early stage of development. Older plants affected by this virus will be further damaged as the leaves die back from the tip downwards.

Treatment: This is a difficult virus for growers to manage, as it can overwinter in nearby grasses, and it also affects most cereal crops, so rotation to another cereal will not cause its depletion in the field. The disease quickly spreads via both local and migratory aphid vectors. As a result, infection can occur from the early growth stages right up to maturity.

Figure 8.3 Comparison between a healthy barley plant (shown on left) and a barley plant infected with barley yellow dwarf virus (BYDV) (shown on right).

8.3 Beet Leaf Curl Virus (BLCV)

Family: Rhabdoviridae

Genus: *Nucleorhabdovirus*

Main hosts: Fodder beets and sugar beets

Structure: ssRNA–, linear, bacilliform, enveloped, large (often over 200 nm in length)

Transmission: Via beet lace bug (*Piesma quadratum*)

The Rhabdoviridae are a large family of viruses that infect both plants and animals (famously, the rabies virus belongs to this group). Plant virus genera in this family include *Cytorhabdovirus* (these species replicate in the cytoplasm) and *Nucleorhabdovirus* (these species replicate in the nucleus). BLCV belongs to the latter genus and is found mainly in mid and eastern Europe. The virus affects the leaves of the beet plants by causing deformation of the veins, which leads to the typical curling of the leaf structure (Figure 8.4); it also causes bleaching of the midrib and dwarfing of the whole plant, including the roots. Sugar content in the beets can be severely impaired by this virus, reducing their value as a crop. Outbreaks of the disease tend to be sporadic, correlating with changes in the vector population.

Treatment: Diseased plants cannot be treated, so prevention is the only form of control. Targeting the vectors with pesticides, using more resistant host varieties, and implementing crop rotations can be effective. The virus has a limited host range, so using non-host plants for at least 2–3 years is advised. Good management practices in recent years have led to this disease becoming rare.

Figure 8.4 Beet leaf curl virus (BLCV) on a leaf of the common beet (*Beta vulgaris*). (Courtesy of Pflanzenschutzamt, Hanover Archive, Bugwood.org)

8.4 Cauliflower Mosaic Virus (CaMV)

Family: Caulimoviridae
Genus: *Caulimovirus*
Main hosts: Cauliflower and other brassicas
Structure: dsDNA (RT), icosahedral
Transmission: Via aphids

CaMV typically infects brassicas, including cauliflower (after which it was named), as well as a number of other family groups, including the Solanaceae. It is unusual for a plant virus in that it replicates by **reverse transcription (RT)**. Symptoms include mosaic effects, necrotic lesions, and whole-plant stunting. Transmission is by aphids, but unlike many other aphis-transmitted viruses, CaMV does not enter the vector's circulatory system, but instead attaches to the outer mouthparts (the stylets), assisted by a virus helper protein that increases infection rates in new hosts. Recent research indicates that the virus utilizes the host's own cellular pathways to facilitate this process.

8.5 Color-Breaking Potyviruses

Family: Potyviridae
Genus: *Potyvirus*
Main hosts: Tulips, lilies
Structure: ssRNA+, filamentous
Transmission: Via insects (mainly aphids)

The effect of the color-break viruses in tulips was discussed briefly in Chapter 3, and its effect on this host is illustrated in Figure 3.16. Here the potyvirus interferes with anthocyanin pigment distribution in the flower petals, causing distinctive regions to become clear of color. To date, five different strains of potyvirus have been identified that cause color break in both tulips and lilies, although there are no obvious specific differences in their effects, and the strains can only be distinguished using serological tests and molecular analysis. These five strains are tulip-breaking virus (TBV), tulip top-breaking virus (TTBV), tulip band-breaking virus (TBBV), lily mottle virus (LMoV), and Rembrandt tulip-breaking virus (ReTBV). Although only tulips and lilies are affected by these color-breaking potyviruses, it should be noted that other color-breaking viruses from other virus families affect other flowering plants, such as roses and orchids; these viruses are less common and much less widely documented.

Treatment: Although there is no treatment for this virus, and indeed in tulips it is considered useful for inducing desirable effects, it is advisable to avoid planting lilies close to tulips that have been bred to include the virus.

8.6 Grapevine Vitiviruses A, B, D, E, and F

Family: Betaflexivirideae
Genus: *Vitivirus*
Main hosts: Grapevines (*Vitis* species)
Structure: ssRNA+, filamentous
Transmission: Via aphids, mealy bugs, and propagation techniques

This is a small group of type IV filamentous plant viruses that affect grapevines, reducing plant vigor and resulting in significantly lower crop yields. Symptoms include reddening of the leaf margins and petioles, leafroll, swelling at the graft union, and reduced plant size. Corky wood is a further symptom typically arising in infections with grapevine vitivirus B (a strain that only appears to infect grafted specimens). Transmission is by insects and by poor hygiene during handling. Increasing hygiene of both hands and tools during propagation reduces the risk of infection.

8.7 Helleborus Net Necrosis Virus (HeNNV)

Family: Betaflexivirideae
Genus: *Carlavirus*
Main hosts: Hellebore (*Helleborus* species)
Structure: dsRNA
Transmission: Via aphids and leafhoppers

HeNNV is a fairly recently identified virus, having been observed only since the 1990s in Europe, since around 2000 in the USA, and even more recently in Japan and New Zealand. Its symptoms are similar to those of a number of fungal pathogens, but the additional symptoms of mottling associated with the leaf veins, together with black streaking starting at the leaf veins and progressing down the petiole and stems, are characteristic of this virus. Black streaks may also appear on the sepals and carpels. Other symptoms include stunted and distorted growth.

Treatment: There is no effective treatment for this disease. Infected plants should be removed and destroyed.

8.8 Peanut Clump Virus (PCV) and Indian Peanut Clump Virus (IPCV)

Family: Virgaviridae
Genus: *Pecluvirus*
Main hosts: Peanut (*Arachis hypogaea*) and other legumes, and members of many other families
Structure: ssRNA+, non-enveloped rod
Transmission: Via seed, soil fungi, and protists

This disease relationship is full of misnomers! The peanut (also known as groundnut) is not a true nut, but a legume. The virus infects not just the peanut, but also many other legumes and a wide range of other family groups, including cereals, sugar cane, and cucurbits. Symptoms include clumping and also red leaf mottling and ringspots.

PCV is found mainly in West Africa, where peanut is an important cash crop. Yield losses can be as much as 60% following infection. Transmission is via host plant seed and soil fungal vectors, and PCV is one of a number of soil-borne viruses that are transmitted by the root endoparasite vector *Polymyxa graminis*, which is a protist. A similar disease with similar symptoms has been reported in India and is known as Indian peanut clump virus (IPCV). However, PCV and IPCV are not thought to be related. A clump virus has also recently been reported in China, although this has yet to be characterized.

Treatment: As a crop that grows in areas with high levels of abiotic stresses (for example, unpredictably high temperatures, long periods of drought), peanut plants are rendered vulnerable to disease infections, and in this case we see that even the vectors are infectious agents. Virus infections in plants are not treatable, so prevention is the only method of control. Crop rotations can be partially effective, although the broad range of transmission types reduces this effect. Some research has been undertaken to genetically modify the coat proteins with a view to increasing resistance to viral disease.

8.9 Plum Pox Virus

Family: Potyviridae
Genus: *Potyvirus*
Common name: Sharka
Main hosts: Drupes (stone fruits)
Structure: ssRNA+, filamentous
Transmission: Via aphids

Commonly known as sharka (which is Slavic for pox), the plum pox virus (PPV) is probably the most commercially significant and widespread virus of the **drupes** (stone fruits). All parts of the plant are affected. Discolored rings develop on the surface of the leaves, and may also develop on the skin of infected fruits. Infected fruits may be small, deformed, and can develop a tart taste due to increased acidity and reduced sugar production.

Treatment: This virus is transmitted by aphids and is therefore very difficult to control. In an area where on outbreak occurs, all of the infected fruit trees have to be destroyed in order to prevent spread of the disease. Research is currently focusing on disease resistance as a way forward for the industry. Some natural resistance has been observed, giving plant breeders the opportunity to make use of the resistance traits in these plants to develop new, more resistant fruit tree cultivars for the market. In addition, a new transgenic variety of plum called "Honeysweet" has been developed by utilizing the virus's own capsid protein gene to transform the plant. These experimental trees have shown notable resistance in field trials.

8.10 Potato Virus Y (PVY)

Family: Potyviridae
Genus: *Potyvirus*
Main hosts: Members of the family Solanaceae, and many other species
Structure: ssRNA+, filamentous
Transmission: Via aphids (often with helper viruses)

This virus gives its name to the whole family of viruses (the Potyviridae), although PVY is itself a distinct species from many of the other viruses in the family. Transmission is mainly by aphids, and is often facilitated by helper viruses. This transmission is classed as non-persistent (that is, the virus is not perpetuated in the aphid's body, but simply becomes attached to the vector's stylet and is then transmitted to the next plant when the aphid moves on). Because of this, the transmission efficiency for any one aphid is poor. However, aphids exist in vast numbers, increasing the likelihood of infection within the crop.

PVY has many strains and affects a number of hosts in a variety of ways, depending on which virus strain, which host species, and (in the case of crop plants such as potatoes) which variety of host is involved. Environmental and temporal factors also influence the severity of the infection. Here we discuss the potato as the main host, but this virus also infects other *Solanum* species, such as tomatoes, peppers, and aubergines, as well as some horticulturally important plants, such as the common summer bedding plant, petunia.

Mild cases of this disease cause stunted growth and mottling of leaves, which decreases photosynthetic efficiency and thus reduces crop yields. However, in more severe cases of the disease, infection may cause potato tuber necrotic ringspot disease (**Figure 8.5**).

Figure 8.5 Potato showing potato tuber necrotic ringspot disease caused by potato virus Y (PVY). (Courtesy of Graeme Stroud, Agriculture and Horticulture Development Board.)

This is a well-known and well-studied virus, and consequently many cultivars of potato have been bred with resistance in mind. In addition, growers are advised by agricultural governing bodies (for example, the Potato Council and Defra in the UK, the USDA in America, and many others across Europe) about where to source their seed potatoes in order to obtain clean stock and avoid spread of the disease.

8.11 Raspberry Ringspot Virus (RRSV) (also known as RpRSV)

Family: Secoviridae (subfamily Comovirinae)
Genus: *Nepovirus*
Main host: European red raspberry (*Rubus idaeus*)
Structure: ssRNA+, non-enveloped, icosahedral
Transmission: Via nematodes and seed

Raspberry ringspot virus (RRSV) was first identified in Scotland in the 1950s, and has since been recorded throughout most of Europe and parts of Russia. It infects mainly raspberries (**Figure 8.6**), but has also been recorded in other *Rubus* species, berry fruits, currants, and cherries, as well as historically causing some significant economic problems in grape crops in Germany.

Treatment: The genus *Nepovirus* by definition contains the nematode-transmitted viruses, and consequently fumigant systems have been used to control the nematode vectors with some success. However, the disease is also transmitted via seed, so controls using certified RRSV-free plant stock are utilized commercially, and indeed small fruit crops are generally propagated vegetatively. Very little research on this virus is currently ongoing, largely due to the success of commercial control systems. However, this disease also infects wild species, so it is likely that over time new strains of the virus will arise and may be transmitted to commercial fruits.

8.12 Rice Tungro Spherical Virus (RTSV)

Family: Sequiviridae
Genus: *Waikavirus*
Main hosts: Rice, and some other grass species
Structure: ssRNA+, spherical
Transmission: Via aphids and leafhoppers

Figure 8.6 Leaves of European red raspberry (*Rubus idaeus*) showing characteristic damage caused by raspberry ringspot virus (RRSV). (Courtesy of SCRI-Dundee Archive, Scottish Crop Research Institute, Bugwood.org)

Rice forms the staple diet of literally billions of people around the world, and is a dietary choice for many others. It is a global crop, grown across the tropics, subtropics, and also some temperate areas. However, around 75% of all rice is grown in the south and eastern regions of Asia. It is here that rice tungro spherical virus (RTSV) (also known as tungro or waika) is the most significant viral threat to cropping systems, although rice is affected by at least 30 different viruses from a range of families.

RTSV is unusual in that it commonly forms an unlikely alliance with another virus, the rice tungro bacilliform virus (RTBV)—a dsDNA virus—significantly increasing its success rate. It is noteworthy that when a complex between the two viruses is formed, only the green leafhopper transmits the disease.

Tackling the vector is one major method of control, although this is costly and rice fields cover vast areas. Much research is currently being undertaken on the use of transgenic control methods. For example, Tyagi et al. (2008) have been developing RNA interference techniques by producing transgenic rice plants that express dsRNA. By altering the genetic makeup of the host plants, a decrease in virus accumulation in the host plants was shown to occur, and symptoms were subsequently reduced.

8.13 Tobacco Mosaic Virus (TMV)

Family: Unassigned
Genus: *Tobamovirus*
Main hosts: Members of the family Solanaceae, and many ornamental plants
Structure: ssRNA+, helical
Transmission: Via insect vectors and contact with other plants

TMV was the first virus to be identified, and is possibly the most studied virus to date, with much of our current knowledge about viruses generally having been obtained by studies and documentation of the structure and functions of TMV. Its capacity to self-assemble *in vitro* has made it an ideal candidate for molecular analysis. It consists of coiled ssRNA+ encased in a helical arrangement of protein subunits that form the capsid for this rod-shaped virus.

TMV was named for the host plant—tobacco—in which it was first identified as an infectious agent, and for its morphological effect on the host plant, which causes a mottling mosaic appearance. However, it has a huge host range, mainly consisting of members of the family Solanaceae, but also including cucurbits and many ornamental plants, such as the sunflower (Figure 8.7). Crop yields are not greatly affected by TMV, although even the small percentage of losses for high-value tobacco crops can be economically significant, possibly reducing the gross national product (GNP) of areas such as North Carolina in the USA.

Life Cycle: TMV is a remarkably resilient virus, capable of surviving for long periods on minimal infected tissues, and tolerant of high temperatures and dry conditions. It overwinters in infected debris from its host, and then infects new hosts, either by direct contact or by vector transmission, usually via aphids and leafhoppers. Successful infection involves passage through the host tissues from cell to cell via the plasmodesmata (narrow pores connecting adjacent plant cells that enable transport of materials from cell to cell), and like a number of other viruses it facilitates this process by producing a movement protein that enlarges these access points.

Treatment: As with all viruses, there is very little that can be done to treat infected plants. However, prevention measures can be a very effective form

---

8.2 From Science Fiction to Science Fact

The unique properties of tobacco mosaic virus (TMV) have attracted the attention of microengineering technologists, and significant research is currently focusing on the uses of modified TMV for the production of nanowires that can be utilized for nanostructure material synthesis. In addition, it has been shown that TMV enhances the efficiency of lithium-ion microbatteries by almost an order of magnitude.

(a)

(c)

(b)

Figure 8.7 Tobacco mosaic virus (TMV) on sunflower. (a) Collection of sunflowers: note that the specimen on the right is much smaller, in terms of both height and leaf and bloom size, due to infection by TMV; (b) the infected sunflower shown in more detail; (c) one of the leaves of the infected sunflower—note its small size and distorted, leathery, and mottled appearance. A color version of this figure can be found in the color plate section at the end of the book.

of control. They include good hygiene to reduce transmission of infection, implementation of crop rotations with non-host plants, use of more resistant crop plant cultivars, and also possibly inoculations of attenuated TMV to outcompete the virulent strains.

8.14 Tobacco Ringspot Virus (TRSV)

Family: Secoviridae

Genus: *Nepovirus*

Main hosts: Tobacco (*Nicotiana tabacum*), soya bean (*Glycine max*), and many others

Structure: ssRNA+, icosahedral

Transmission: Via nematodes, mites, thrips, and seed

Like TMV, TRSV is a broad-ranging virus that affects numerous hosts and causes a wide range of symptoms. In tobacco plants, characteristic ringspots develop (Figure 8.8), giving the disease its name. In severe infections this can lead to necrotic lesions. Soya bean is another important crop affected by TRSV, which in this species typically causes bud blight and growth stunting. Transmission is via nematodes, a number of insect vectors, and host seed.

Treatment: Use of clean seed stock and resistant plant varieties may prevent infection. However, the wide host range makes this disease difficult to control.

8.15 Tomato Bushy Stunt Virus (TBSV)

Family: Tombusviridae

Genus: *Tombusvirus*

Main hosts: Tomato (*Solanum lycopersicum*, syn. *Lycopersicum esculentum*) and other members of the family Solanaceae

Structure: ssRNA+, non-enveloped, icosahedral

Transmission: Unknown

8.3 A Plant Virus That is Also Pathogenic to Bees

It has recently been reported that tobacco ringspot virus (TRSV) has made the unusual leap from the plant kingdom to the animal kingdom, and now affects honey bees. Bees have been implicated as vectors of the disease via pollen transmission, but now it seems that they too are adversely affected by the virus. In normal circumstances, insects that function as plant virus vectors carry the disease agent in their alimentary canal and/or their salivary glands, where the virus may or may not replicate. However, in the case of TRSV in honey bees, virus particles have been found throughout the body of the bee, including the wings, antennae, and nervous system. As a result, TRSV has been implicated as one of the causes of bee colony collapse.

Figure 8.8 Burley tobacco leaf (*Nicotiana tabacum* burley type) showing clear symptoms of tobacco ringspot virus (TRSV). (Courtesy of Paul Bachi, University of Kentucky Research & Education Center, Bugwood.org)

TBSV typically infects members of the family Solanaceae, including tomatoes, from which the disease gets its name. It has been identified in Europe, throughout the Americas, the Middle East, and parts of Northern Africa. Symptoms include chlorotic lesions, necrotic lesions in more extreme cases, mosaic effects, reduced leaf size, reduced or absent fruits, and proliferation of lateral shoots—giving the characteristic "bushy" appearance.

This is generally regarded as a soilborne virus, although no transmission methods have been identified to date.

Treatment: A number of resistant tomato varieties have been identified. The implementation of crop rotations is also advisable.

Note: Research indicates that salicylic acid (SA), in addition to its well-known role of engendering resistance to viruses, may have a post-infection role of inhibiting TBSV replication by targeting the host component in the virus–host replication complex.

Further Reading

Abo ME & Sy AA (1998) Rice virus diseases: epidemiology and management strategies. *J Sustain Agr* 11:113–134.

Agrios GN (2005) Plant Pathology, 5th ed. Elsevier Academic Press.

Dekker EL, Derks AFLM, Asjes CJ et al. (1993) Characterisation of potyviruses from tulip and lily which cause flower-breaking. *J Gen Virol* 74:881–887.

Eastwell KC, du Toit L & Druffel KL (2009) Helleborus net necrosis virus: a new *Carlavirus* associated with 'Black Death' of *Helleborus* spp. *Plant Dis* 93:332–338.

Garcia JA, Glasa M, Cambra M & Candresse T (2014) *Plum pox virus* and sharka: a model potyvirus and a major disease. *Mol Plant Pathol* 3:226–241.

Hull R, Covey SN & Dale P (2000) Genetically modified plants and the 35S promoter: assessing the risks and enhancing the debate. *Microb Ecol Health Dis* 12:1–5.

Jackson AO, Dietzgen RG, Goodin MM et al. (2005) Biology of plant rhabdoviruses. *Annu Rev Phytopathol* 43:623–660.

Krishna G, Singh BK, Kim E-K et al. (2015) Progress in genetic engineering of peanut (*Arachis hypogaea* L.)—A review. *Plant Biotechnol J* 13:147–162.

Li JL, Cornman RS, Evans JD et al. (2014) Systemic spread and propagation of a plant-pathogenic virus in European honey bees, *Apis mellifera*. *mBio* 5:e00898-13.

Martinière A, Zancarini A & Drucker M (2009) Aphid transmission of *cauliflower mosaic virus*: the role of the host plant. *Plant Signal Behav* 4:548–550.

Nawaz HH, Umer M, Bano S et al. (2014) A research review on tomato bushy stunt virus disease complex. *J Nat Sci Res* 4:18–23.

Pomerantseva E, Geraspoulos K, Gnerlich M et al. (2013) Creating High-Performance Microbatteries with Tobacco Mosaic Virus. *SPIE Nanotechnology* (doi:10.1117/2.1201309.005050).

Tian M, Sasvari Z, Gonzalez PA et al. (2015) Salicylic acid inhibits the replication of *tomato bushy stunt virus* by directly targeting a host component in the replication complex. *Mol Plant Microbe Interact* 28:379–386.

Tyagi H, Rajasubramanian S, Rajam MV & Dasgupta I (2008) RNA-interference in rice against Rice tungro bacilliform virus results in its decreased accumulation in inoculated rice plants. *Transgenic Res* 17:897–904.

Useful Websites

CABI. Plantwise. www.plantwise.org

Commonwealth Scientific and Industrial Research Organisation (CSIRO). www.csiro.au

Chapter 9
Other Diseases

There are numerous diseases which are attributed to pathogens that do not fit into the taxonomic "big three" (the fungi, bacteria, and viruses), so in this chapter we highlight a diverse range of disease profiles aimed at covering some of these examples.

Many such diseases teeter on the edge of the other groups. For example, the Oomycota or water molds can cause serious problems in a wide range of plants, including many of the economically important crops. There are a number of important genera in this group, such as the *Pseudoperonospora* species, which cause downy mildew in a variety of plants (for example, cucurbits, hops, and hemp), and the *Phytophthora* species, such as *P. infestans*, which has historical significance (as the cause of the Irish Potato Famine in the nineteenth century) (Figure 9.1), and other *Phytophthora* species that have been much in the news in recent years, particularly with regard to the many tree diseases that are spreading throughout Europe and many other parts of the world (for example, *P. ramorum*). However, this group is ubiquitous in that it consists of hundreds of species and infects such a wide host range (for example, *P. megakarya* infects cocoa (*Theobroma cacao*) plants (Figure 9.2).

Figure 9.1 Field-scale image of potato crop infected with *Phytophthora infestans*. A color version of this figure can be found in the color plate section at the end of the book.

Figure 9.2 *Phytophthora* **crown rot on pod of cacao (***Theobroma cacao***).** A color version of this figure can be found in the color plate section at the end of the book. (Courtesy of Christopher J Saunders, Bugwood.org)

9.1 Lichen Pathogenic on Trees

Cephaleuros virescens sometimes constitutes the algal part of a group of lichens called *Strigula fries*. This is the only lichen group currently known to be parasitic on plants (although this is not always the case!). The fungal component of these lichens can belong to any of various genera, but the parasitic lichen typically comprises fungi of the genus *Massaria*, which has recently become infamous as the causal agent of London plane dieback disease.

This chapter will also consider diseases caused by algae (different algal species belong to different kingdoms) and phytoplasmids (belonging to the domain Bacteria, but quite different to the Eubacteria, otherwise known as the true bacteria). For the sake of completeness, nematodes are included in this chapter. Although they are not strictly speaking pathogens, nematodes do cause disease symptoms in plants (for example, potato white cyst disease), so should be considered here for reasons of comparison.

9.1 *Cephaleuros virescens*

Descriptor: Green alga (Chlorophyta)
Family: Trentepohliaceae
Genus : *Cephaleuros*
Main hosts: Many tropical and subtropical plants, including coffee and tea
Common names: Algal leaf spot, green scurf, or red rust (when infecting tea or coffee plants)

There are a number of *Cephaleuros* species that parasitize higher plants, *C. virescens* being one of the most well known. Although it can sometimes spread to the fruits and the stems, this pathogen typically infects leaves by penetrating both the upper and lower epidermal tissue (**Figure 9.3**), causing necrosis and lesions. This pathogen is generally non-lethal, but the amount of damage caused to the host varies according to host species or cultivar, plant health, and immediate environmental conditions. The pathogen can often cause considerable losses to commercial crops such as coffee, tea, guava, and mango.

Life Cycle: *Cephaleuros virescens* is a filamentous alga that thrives in warm, moist conditions as it requires a film of water to complete its life cycle. The zoospores (asexual spores) germinate on the leaf surface when moisture levels are high, and the thalli then penetrate the leaf epidermis and grow within the host plant tissues. The following year, these thalli produce sporangiophores and spherical gametangia (that is, the pathogen exhibits alternation of generations). The resulting sporophytes are sexual structures, although in the case of *C. virescens* these are diminutive, and are a less important stage than the growth of the thalli in terms of causing damage to the plant host. To complete the cycle, spores can be transmitted from plant to plant via water splash.

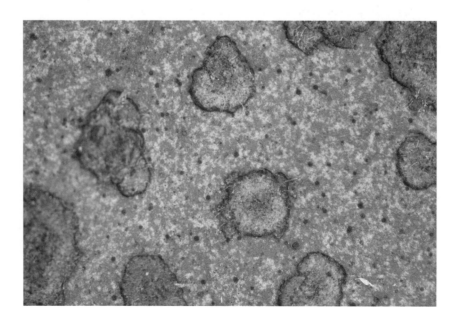

Figure 9.3 Citrus leaf showing the typical penetrating lesions of the parasitic alga *Cephaleuros virescens.* (Courtesy of Cesar Calderon, USDA APHIS PPQ)

Treatment: Maintaining the plant hosts in favorable conditions by reducing stress factors such as poor soils and poor drainage helps them to overcome this pathogen. Resistant cultivars of many crops are also available on the market.

9.2 *Globodera pallida*

Descriptor: Nematode
Phylum: Nematoda
Genus: *Globodera*
Main hosts: Members of the family Solanaceae, especially potatoes
Common name: White potato cyst nematode

Nematodes are the only animal pathogen included in this book. Strictly speaking, nematodes are classed as parasites, although the symptoms that appear on the host plants are pathogen-like, so this organism has been included here for the sake of completeness.

It is not known how many nematode species exist, although the likely figure is considered to be in the hundreds of thousands. A vast number of these are parasitic, while others are carnivorous and are predatory on other agricultural pests. Therefore, depending on the species, these organisms may have a positive or a negative impact on commercially grown plants. Nematodes are globally ubiquitous, having been found in all terrestrial and aquatic environments.

In this chapter we highlight a single species, the white potato cyst nematode (*Globodera pallida*), which is one of the most important parasites of potato crops across many countries, including most of Europe, Asia, and the Americas.

Life Cycle: As the common name of this species implies, white cysts form on the roots of infected potato plants (**Figure 9.4**). However, these structures are not symptoms of the disease, but rather they are the dead bodies of the female nematodes. These tough structures are attached to the host plant roots, and the eggs of the females remain inside their bodies until they hatch. After hatching, the juveniles (second-stage juveniles) are in the infectious stage of their life cycle and penetrate the root cells, where they insert their stylet (**Figure 9.5**) into a selected cell and induce development of a **syncytium** (a proliferation of multinucleate merged cells), where higher nutrient levels will be provided by the host. In the third stage of their life cycle, those that penetrate the pericycle cells typically become female, whereas those that penetrate the cambium cells typically become male, and the life cycle begins again. Symptoms on the host plant include yellowing, wilting, dwarfing, and early senescence.

Treatment: These nematodes can be difficult to identify and to treat. However, their ubiquitous nature indicates that constant precautions are needed with regard to hygiene, keeping machinery cleared of soil, use of clean water, and implementing crop rotations.

Figure 9.4 The small white cysts that are typical of the nematode parasite *Globodera pallida*. (Courtesy of Bonsak Hammeraas, Norwegian Institute of Bioeconomy Research, Bugwood.org)

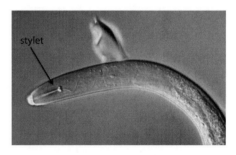

Figure 9.5 The nematode parasite *Globodera pallida* in the second stage of its life cycle. Note the distinctive stylet extending from the mouth. (Courtesy of Christopher Hogger, Swiss Federal Research Station for Agroecology and Agriculture, Bugwood.org)

9.3 Peach X-Disease Phytoplasma

Descriptor: Tenericute (part of the domain Bacteria)
Family: Mycoplasmataceae
Genus: *Phytoplasma*
Main hosts: Stone fruits (drupes), including peaches, nectarines, plums, and cherries
Alternative names: Western X-disease, X-disease

Figure 9.6 Leaves of a peach plant (*Prunus persica*) infected by peach X-disease phytoplasma. Wilting and curling of the leaves, together with lesions along the leaf edges, are clearly visible. (Courtesy of F Dobsa, INRA, Bordeaux, Bugwood.org)

This disease is severe, although not typically common. After infection, the plant will be asymptomatic for at least 1 year and often up to 2 years while the pathogen establishes itself, which makes diagnosis difficult. When symptoms do appear, the leaves start to yellow, curl, and develop lesions along the edges (Figure 9.6), together with spotting across the leaf. Leaves and fruits drop prematurely, reducing crop yields.

Life Cycle: This pathogen typically resides in the phloem tissue of the host plant, and is spread through the crop by means of insect vectors, usually leafhoppers. Replication is by binary fission, and the long incubation period allows substantial colonies to develop.

Treatment: There is currently no treatment for the disease, although control of vectors with insecticides is effective. Infected trees must be removed and burned to prevent infection of crops in later years.

9.4 *Phytomonas staheli* and *Phytomonas leptovasorum*

Descriptor: Trypanosomatid flagellate
Family: Trypanosomatidae
Genus : *Phytomonas*
Main hosts: Coconut palm, oil palm, coffee
Common names: Sudden wilt (in coconut palm); hartrot (in coconut palm and oil palm); slow wilt, lethal yellowing, and bronze leaf wilt (in oil palm); phloem necrosis disease (in coffee)
Transmission: Via insect vectors (mainly Hemiptera; for example, spurge beetles)

Members of the family Trypanosomatidae are uniflagellate, unicellular, eukaryotic organisms, and are represented by a substantial number of well-known parasites that affect humans (for example, sleeping sickness) and animal livestock. However, within this family the genus *Phytomonas* is associated with plants and insects. Most members of this genus are endophytic (non-parasitic). However, two species have been demonstrated to be parasitic on plants, namely *P. staheli*, which causes wilt symptoms in coconut palm and oil palm, and *P. leptovasorum*, which causes phloem necrosis disease in coffee. Both of these parasitic species have been isolated in South America, and *P. staheli* has also been isolated in Malaysia, although *Phytomonas* species have a global distribution, so other parasites may be identified in the future in other regions.

There have been a number of reports of possible parasitism by other *Phytomonas* species in other host species, but to date these have not been confirmed. Indeed the genus was first identified in the latex of *Euphorbia pilulifera*, and although claims have been made that the organism reduces starch granules in the latex, there is no evidence to support this theory. A similar example with regard to research into other diseases of host plants relates to tomato and cassava plants, which have been implicated as hosts, although to date this has not been confirmed.

Life Cycle: The life cycle of *Phytomonas* has not yet been clearly established, but it appears to be reliant on, and linked to, the transmission of this pathogen via sap-sucking insect vectors. Therefore, as in other members of the family, life-cycle stages would occur both in the plant host tissue and in the alimentary canal of the vector.

Treatment: There is no known treatment at present, and crops must be removed and destroyed if they become infected. Use of insecticides to control vectors is one possible approach, although these insects are widespread and populations would re-establish quickly.

9.5 *Phytophthora ramorum*

Descriptor: Oomycote (water mold)

Family: Pythiaceae

Genus: *Phytophthora*

Main hosts: Mainly trees and shrubs, especially oak, larch, chestnut, rhododendron, and bilberry

Common names: Sudden oak death, dieback disease, ramorum disease, larch disease

It is not known how many species of *Phytophthora* exist, but estimates are in the hundreds. *P. ramorum* has been highly publicized on account of its infection of oak trees (*Quercus* species), causing sudden oak death, although oak trees in the UK have proved to be more resistant than oak species in America (where the disease is thought to have originated). However, larch trees (*Larix* species) have been severely affected, and felling rates, particularly in western parts of the UK, have been high. The Forestry Commission in the UK has carried out extensive aerial surveys to obtain data on the spread of the disease. It is then able to use diagnostic kits called lateral flow devices to positively identify the presence of this disease.

In 2015, *P. ramorum* was identified in sweet chestnut (*Castanea sativa*) trees at Duchy College, Stoke Climsland, Cornwall. The college complied with Forestry Commission directives to remove and destroy a number of trees, and to monitor a number of others. Before these were removed, the author was able to collect samples from some of these trees (**Figure 9.7**) and analyze the disease using scanning electron microscopy (SEM) (**Figure 9.8**).

(a)

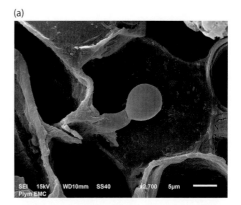

(b)

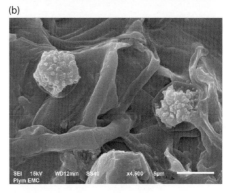

Figure 9.7 *Phytophthora ramorum* **infection of sweet chestnut (***Castanea sativa***).** A color version of this figure can be found in the color plate section at the end of the book.

Figure 9.8 Scanning electron microscopy (SEM) images of sweet chestnut (*Castanea sativa*) infected with *Phytophthora ramorum.* (a) Developing oogonium attached to mycelia; (b) developing spores.

Figure 9.9 Schematic diagram of the life cycle of the family Pythiaceae using a generalized plant host indicating both sexual and asexual reproduction. A color version of this figure can be found in the color plate section at the end of the book.

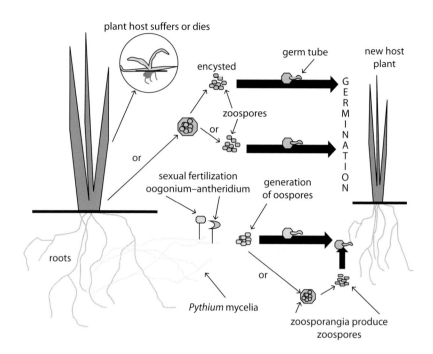

Life Cycle: *Phytophthora* belongs to the same family as *Pythium* and has a similar life cycle to the latter, with numerous strategies for spore production and thus propagation (Figure 9.9).

Treatment: There is no known treatment for *Phytophthora ramorum* (or indeed for other diseases caused by *Phytophthora* species). Strict hygiene measures help to reduce potential infection, but when the disease is found in sporulating host trees (that is, trees in which the pathogen can produce spores), those trees have to be felled and destroyed. There are also a number of recorded instances of infection of terminal hosts (for example, the Douglas fir), in which the pathogen does not reproduce. Although this may be regarded as less serious, it is worryingly indicative of the disease's broad spectrum of infectivity.

9.6 *Phytoplasma asteris*

Descriptor: Tenericute (part of the domain Bacteria)
Family: Mycoplasmataceae
Genus: *Phytoplasma*
Main hosts: Broad host range, including asters and other members of the family Asteraceae, as well as carrots and grapevine
Common name: Aster yellows
Transmission: Via leafhopper vectors, such as the aster leafhopper (*Macrosteles quadrilineatus*)

This disease is named for its effects on asters (members of the family Asteraceae), in which it was first identified. However, it has a vast range of hosts, including many cereal crops, ornamental plants, and vegetables, most notably carrots (*Daucus carota*). *P. asteris* has a global distribution, and therefore its economic importance cannot be overstated. Symptoms of the disease include yellowing (chlorosis) of leaves, vein clearing, stunting of the main roots, and proliferation of adventitious roots (Figure 2.14).

Life Cycle: As is typical of mycoplasmids, replication is mainly by binary fission, although it can also occur by budding. This occurs both in the plant host phloem tissue and in the alimentary canal of the leafhopper vector. Recent research on infected grapevines in Italy indicates that the leafhoppers preferentially select infected plants to feed on, possibly due to the yellow coloring acting as an attractant, thereby perpetuating the success of the pathogen.

Treatment: There is no known treatment for this disease, and infected plants must be removed and destroyed. The use of insecticides to control the vector may have some impact.

9.7 Pseudoperonospora humuli

Descriptor: Oomycete (water mold)
Family: Peronosporaceae
Genus: *Pseudoperonospora*
Main host: Hops
Common name: Hop downy mildew

The growing of hops (*Humulus lupulus*) in the UK is commonly associated with Kent (the Garden of England) and Sussex. However, this is a global crop that is not only grown extensively in Europe, but also cultivated in Asia, the Americas, Africa, and Australasia—due largely to the ubiquitous popularity of beer, with hops being one of the main ingredients of traditional ale. There is an extensive cultural history associated with the growing and harvesting of this particular crop. For example, the skilled use of stilts, originally for maintenance of the hop **bines** (**Figure 9.10**) as well as that of some other tall crops, later became popular at county fairs and parades. The large-scale temporary migration of Londoners to the Kent countryside at harvest time may have been a precursor to the "common folk" of southern Britain taking country and coastal holidays—a concept previously associated with the gentry!

There are many diseases that infect hops. A number of them are exclusive to this plant and, of these, none are more devastating than *Pseudoperonospora humuli*, which if left unchecked will cause significant damage to all of the aerial parts of the plant. (*P. humuli* is closely related to *P. cubensis*, which is a common infectious agent of cucurbits.)

Life Cycle: *P. humuli* is transmitted from plant to plant via wind and water splash. Zoospores develop rapidly from sporangia following infection of leaves, crown, buds, shoots, and cones. They then engage in a continuous replication cycle as mycelia grow throughout the tissues and sporangiophores emerge, mainly through the stomata, and release new zoospores. At the end of the growth season, mycelia overwinter in the crowns and buds, protected by the host's own cuticle, and sometimes also overwinter in any plant debris left on the ground after harvest. New shoots that are already infected emerge in the following growth season.

Treatment: This is a difficult disease to treat due to its efficient overwintering strategy. Removal of plant debris at the end of the season may prevent some reinfection. Systemic fungicide treatments can be effective against the overwintering mycelia. However, as there is limited plant tissue remaining after harvest this would be insufficient to prevent all infection the following year. Aggressive fungicide treatment early in the season, consisting of several applications, is required to combat this disease. Recent research indicates that different treatments on the market have optimum effectiveness at different stages both pre- and post-harvest. Maintaining a healthy soil profile by mechanical cultivation and manure application may also encourage a range of antagonists to the pathogen.

Figure 9.10 Hop stringing using stilts in Bodiam, Sussex. (Courtesy of Keith Ennis, Bygone Bodiam.)

Figure 9.11 The devastating effect of *Pythium* species on seedlings and cuttings, illustrated by the fatal damage caused to young pepper plants (*Capsicum* species). (Courtesy of Margaret McGrath, Cornell University.)

Figure 9.12 Maize (*Zea mays*) with highly proliferated leafy growth where flowers should be developing. This distinctive growth give this disease its common name, "crazy top disease." (Courtesy of Gregory Shaner, Purdue University.)

9.8 *Pythium* Species

Descriptor: Oomycote (water mold)
Family: Pythiaceae
Genus: *Pythium*
Main hosts: Seedlings and cuttings of most plants
Common name: Damping off disease
Transmission: Self-transmission via motile zoospores, attachment to feet of flying insects (for example, fungus gnats)

Pythium species are oomycotes (water molds) and are closely related to *Phytophthora* species. The genus *Pythium* is a key disease agent in crops, especially those sown in glasshouses, causing damping off of newly germinated seedlings and cuttings (Figure 9.11). There are several hundred species, which infect a wide host range with low host specificity, and they are found globally. The disease can also infect older plants, causing wilt effects by damaging roots, as well as blight of foliage. Together these characteristics result in a disease that is extremely difficult to manage and is therefore of major commercial significance.

Life Cycle: *Pythium* reproduces both sexually via oospores (which may overwinter), and asexually via zoospores. Oogonia and antheridia develop on the mycelia from the infected plant, and then fertilization occurs, resulting in development of the oospores that either germinate directly or further develop into zoosporangia, the spores of which germinate. Alternatively, sporangia may be directly released from the infected plant, and they then either germinate or produce encysted zoospores that later germinate (Figure 9.9). Zoospores require water for motility, so crop infections may be limited to patchy areas in dry conditions.

Treatment: This is a difficult disease to treat because *Pythium* is ubiquitous. During cultivation, strict hygiene measures and disinfection of tools and benches reduce the likelihood of infection, as does the use of filtered irrigation water to remove the motile zoospores and treated compost.

9.9 *Sclerophthora macrospora*

Descriptor: Oomycote (water mold)
Family: Peronosporaceae
Genus: *Sclerophthora*
Main hosts: Cereals, rice, sorghum, and other grasses (including turfgrass)
Common names: Crazy top disease, witches' broom (although the latter name is also used for a number of other diseases and symptoms)

This is an unusual disease that affects the host by causing abnormal changes in tissue that should develop into flowers, so that it develops instead into prolific leafy growth, a phenomenon termed **phyllomorphy** (also known as phyllody). The pathogen infects grasses, especially rice and cereals, including wheat, barley, oats, and maize (Figure 9.12). The disease is easy to identify due to the obvious symptoms that gave rise to its common name, "crazy top disease," but other symptoms can also occur, such as stunting of the whole plant, narrowing of leaves, poor leaf color, and poor seed development. In heads that are phyllomorphically altered, flowers and therefore seeds may be completely compromised. In turfgrass the disease causes stunted yellow tufting, and in sorghum it causes downy mildew—a more typical effect of infections caused by the family Peronosporaceae. The disease has a widespread distribution, being found across Europe, Asia, the

Americas, Africa, and Australasia, although it is favored by damp temperate conditions.

Life Cycle: Like similar pathogens, *S. macrospora* produces both zoospores and oospores to exploit the best opportunities for replication and for infection of new hosts (Figure 9.9).

Treatment: This disease cannot be treated, but strict hygiene measures can reduce the incidence of disease. Most research has focused on the development of more resistant crop cultivars.

Further Reading

Agrios GN (2005) Plant Pathology, 5th ed. Elsevier Academic Press.

Brooks F, Rindi F, Suto Y et al. (2015) The Trentepohliales (Ulvophyceae, Chlorophyta): an unusual algal order and its novel plant pathogen—*Cephaleuros*. *Plant Dis* 99:740–753.

Burchett SG (2005) Powdered seaweed extract of *Ascophyllum nodosum* conveys protection against *Pythium* spp. infection when used as a rooting compound for green and semi-woody cuttings of *Dianthus* and *Cineraria* cultivars. Unpublished report.

Gent DH, Twomey MC, Wolfenbarger SN & Woods JL (2015) Pre- and postinfection activity of fungicides in control of hop downy mildew. *Plant Dis* 99:858–865.

Jaskowska E, Butler C, Preston G & Kelly S (2015) *Phytomonas*: trypanosomatids adapted to plant environments. *PLoS Pathog* 11:e1004484.

Krüger K, Venter F & Schröder ML (2015) First insights into the influence of aster yellows phytoplasmas on the behaviour of the leafhopper *Mgenia fuscovaria*. *Phytopath Moll* 5 (Suppl.):S41–S42.

Useful Websites

CABI. www.cabi.org (a range of up-to-date and relevant information from around the world that students and other readers will find useful)

Forestry Commision. *Phytophthora ramorum*. www.forestry.gov.uk/pramorum

Great Lakes Hops. www.greatlakeshops.com

Section 3
Applied Management of Plant Diseases

Chapter 10: Control of Diseases

This chapter discusses the basic principles of chemical control methods and establishes the concept of reducing the economic threshold of plant pathogens and simultaneously increasing yields. It then discusses the different approaches to control of fungal, bacterial, and viral diseases. The chapter also discusses contact and systemic fungicides (with examples from the major chemical groups, and details of their active ingredients and modes of action), application methods (including the development of spray technology, formulations, and surfactants), precision farming, the development of fungicide resistance, the implications of chemical control methods for the environment, and environmental impact assessment. It then introduces the underlying principles of cultural control and considers the organic philosophy and the establishment of first principles of good husbandry.

Chapter 11: Organic and Integrated Methods

This chapter discusses sanitation and biosecurity, companion planting, applications of element compounds, the crop environment and modifications to that environment, pruning for disease control, the use of rotations, and resistant varieties. It then explores how chemical, biological, and cultural methods can be combined to mitigate against plant disease, by introducing the reader to the underlying principles and focusing on the control of viral and bacterial diseases.

Chapter 12: The Future of Plant Pathology

This chapter explores the state of current knowledge and practices, as well as the future role of plant pathological technology in relation to issues associated with food security and climate change.

Chapter 10
Control of Diseases

The contemporary picture of disease management across the plant production sector is dominated by the agrochemical industry, with images of tractor-mounted sprayers (**Figure 10.1**) ranging across the arable landscape applying a cocktail of crop protection compounds to crops throughout the arable calendar. This perceived dominance of chemicals in the management of crop health also applies to the ornamental plant sector, as a visit to any local garden center will demonstrate—shelves filled with proprietary compounds that are freely available to the gardener. This practice of managing crop health (in this case by controlling disease) is clearly efficient, ensuring a clean and healthy crop throughout the life cycle of the plant, and ultimately maximizing plant quality and the quantity of yield of the harvestable component. To the layman this wide-scale practice may seem uncomplicated and potentially damaging to the environment. Protecting plants from steady and continuous bombardment by an array of different plant pathogens entails a complex mix of art and science. The use of the term "art" in relation to control of plant pathogens is deliberate, and is intended to demonstrate that a degree of judgment is required by the grower, based on the expertise that they may have acquired over a lifetime of growing. Decision making about what and how much pesticide to apply to a crop, and when to do so, is then supported by tools derived from scientific investigation and applied practice.

This chapter aims to introduce the student to the subject of disease management and to the scientific principles that underpin plant health, such as biosecurity, cultural control, plant breeding systems, and disease resistance. Assessments and monitoring of disease, decision-support mechanisms, and the role and application of fungicides are also discussed.

10.1 Principles of Disease Management

There are four principal strategies available to the agricultural and horticultural industries for controlling plant pathogens and pests. These can be summarized as follows:

- **Exclusion of the pathogen from the host.** This includes legislation and quarantine of imported plant material and foodstuffs, and inspection of germplasm and seed certification. Crop inspection forms part of the quarantine and legislation approach, and tree inspections in the UK have increased since 2010 as a result of an increased incidence of numerous threats to trees from fungal and oomycote pathogens. Isolating the pathogen from the host is another component of this strategy, and it can be

Figure 10.1 Tractor-mounted spraying equipment. A color version of this figure can be found in the color plate section at the end of the book.

achieved locally by implementing good biosecurity protocols and cultural controls in the cropping system. Finally, within this strategy, quarantine laws are imposed at points of entry to the UK, and indeed most other countries, with perhaps Australia and New Zealand observing the strictest application of biosecurity protocols at points of entry into their respective countries.

- **Reduction or elimination of the inoculum of pathogens.** When a pathogen has established itself in a cropping system or region, growers can adopt a range of practices that help to reduce the level and incidence of disease inoculum. These practices include crop rotations, sanitation, and soil improvement. Crop rotations have been utilized for several centuries, and the basic principle is to control access to host plants by varying crop species within a production area. Sanitation involves the removal of diseased crop debris (for example, fallen diseased apples). Soil improvement such as drainage can help to reduce the incidence of soilborne pathogens by modifying the soil environment and encouraging root growth and development of crops so that they become less susceptible to disease.

- **Improvement of host resistance.** This strategy has been used since the dawn of agriculture and the domestication of both crops and livestock. With regard to crop species, the application of plant breeding was the cornerstone of the "Second Green Revolution" of the 1960s. Plant breeding has continued to develop and has become a very sophisticated practice that involves gene mapping and understanding the structure and function of disease resistance genes.

- **Protection of the host.** This is the judicious application of either organic or chemical compounds to crops in order to protect them from pathogens. This practice has three underlying principles: (1) the activity of exogenously applied compounds may be based on killing the pathogen on contact; (2) the applied compounds may be systemic in nature, and are transmitted around the host to prevent disease development at sites distant from the point of initial pesticide contact with the host; or (3) application of exogenous compounds may initiate innate host defense mechanisms such as systemic acquired resistance (SAR) (Figure 4.3).

Clearly the first step in effective disease control is simply to ensure that disease propagules do not reach your site. This, in its simplest form, starts at the farm gate with the application of biosecurity, a practice that was once widely adopted across Britain as a result of the War Agricultural Executive Committee (post 1915 and then again post 1938). However, since the "Second Green Revolution" (post 1960) and the widespread application of agrochemicals such as fungicides, biosecurity controls have been poorly implemented across the agricultural sector of the UK until very recently. The mainstream thinking was that we had a range of fungicides with which to control plant health. This was of course a naive stance, both in government and in the industry, because the pathological antagonists quickly adapted to these fungicides and were able to develop resistance to them, thus making control of plant pathogens even more challenging.

Biosecurity starts in the farmyard, or in the potting shed for the keen gardener. Straightforward measures such as cleaning and sterilizing cultivation equipment, secateurs, seed trays, and flowerpots are essential for preventing downstream transmission of pathogens. For example, many pathogens (such as *Fusarium*, *Pythium*, and *Rhizoctonia*) are soilborne or have resting spores that reside in the soil that remains attached to farm cultivating equipment or a gardener's pre-used seed trays and flowerpots. A simple wash in disinfectant and rinse in clean water will reduce the carry-over inoculum of these damaging pathogens. In particular, ornamental and general horticulturalists who propagate plant stock from cuttings and seed in the autumn and winter face a major problem from damping off of seedlings and young plants that grow slowly during the winter season. The use of clean sterilized pots will help to reduce this problem. In recent decades, growers have been neglecting the cultural controls outlined above and relying instead on the widespread use of fungicides, but the recent withdrawal of a number of active compounds by the EU and other national governments, along with the development of fungicide resistance in many pathogens, have seen a resurgence in the use of simple cultural control methods. In the UK this renewed interest has been supported by advice from non-government organizations such as the National Trust (Slawson, 2008), and subsequently from government bodies such as the Forestry Commission (Slawson, 2013).

The next logical step in biosecurity is the selection and use of certified seed and propagules. Traditionally, farmers and growers have saved their own seed and vegetative propagules. However, for decades arable and field-scale horticultural growers have been trading in certified seed, and since 2011 the Seed Marketing (England) Regulations have controlled the trade in certified seed. Under the 2011 Act, seed from the mainstream arable production systems must be certified before it can be marketed. The species or variety has to be listed on a National List or in the EU Common Catalogues. In the UK, the day-to-day technical management of seed certification is overseen by the National Institute of Agricultural Botany (NIAB). The whole process of seed and propagule certification is complex and crop specific, so it is of value in this chapter to review a few major crops, such as the cereals, potato, soft fruit, and top fruit. The rationale for this is to illustrate the fundamental scientific principles that support the food production industry, and to develop the knowledge of the student of plant pathology.

10.2 Seed Certification

The Cereals

All cereal seed traded in the UK must be certified with regard to a number of parameters, including grain quality, viability and germination potential, species and variety purity, percentage of harmful organisms, and (of particular relevance to this chapter) sources of disease inoculum (Table 10.1). Loose

Table 10.1 Seed-borne disease and certification criteria for UK cereals.

Disease		Agronomic Issues		Standards[a]	
Scientific Name	Common Name	Importance	Risk Factors	Regulatory	Advised
Tilletia tritici	Bunt	Low level, but contaminated grain can be rejected	Sowing untreated seed, short rotation between first wheat and second wheat, poor seed bed, dry soil conditions between harvest and sowing	–	1 spore per seed or more
Phaeosphaeria nodorum	*Septoria* seedling blight	High levels of seed-borne inoculum can cause significant loss of young plants at establishment	High inoculum levels in seed, untreated seed sown into poor seed beds, cool and wet soils	–	Treat if over 10% infection of sample
Microdochium nivale and *M. majus*	*Microdochium* seedling blight	A common cause of seedling blight of wheat in the UK, and high levels of seed- and soilborne inoculum will cause substantial loss of young plants at establishment	Wet weather during flowering can lead to high levels of inoculum in seed, sowing untreated seed into poor seed beds, and late sowing of the crop	–	Treat if over 10% infection of sample
Fusarium graminearum	*Fusarium* seedling blight	The only *Fusarium* species to cause significant losses at seedling stage in the UK	High levels of seed-borne inoculum, poor seed bed sanitation, sowing untreated seeds into known infected sites, and including maize in the rotation	–	Treat if over 10% infection of sample
Claviceps purpurea	Ergot	Poisonous both to humans and to livestock. Contaminated grain will be rejected, and minimum standards for the number of ergots per sample exist in certified seed in the UK	Cool wet conditions, grass weeds (which act as host, in particular black grass), prolonged flowering periods, and late tillering	3 pieces per 500 g is the minimum standard, and 1 piece per 1000 g is higher voluntary standard	–
Ustilago nuda f. sp. *tritici*	Loose smut	Not a major issue in the UK, due to seed certification scheme and resistant varieties	Untreated seed, infected neighboring crops, and cool moist conditions during flowering	0.5% infection in embryos is minimum standard, and 0.2% infection in embryos is higher voluntary standard	–
Cochliobolus sativus	Seedling blight and foot rot	A disease of hotter climates so currently not a major problem in the UK, but may become an issue if the climate of the growing season in the UK continues to become warmer	Extended periods of warm moist weather, slow germination and establishment because of poor seed beds	–	–

[a]Standards are either statutory or advisory with regard to certification failure and/or seed treatment.

smut and ergots have minimal quality standards, but no statutory standards exist for bunt and the seedling blights caused by *Microdochium* species and *Septoria* species. There are several categories of seed certification, depending on where the variety is positioned within the seed production system, and each category has specific certification criteria (Table 10.2). Cereal seed in the UK and across Europe must undergo a series of trials to establish variety traits and agronomic performance. In the UK, this is often carried out

Table 10.2 Cereal seed categories and their associated quality standards with regard to viability and purity.

Seed Category	Description	Minimum Germination (% of Pure Seed)	Minimum Analytical Purity (% by Weight)
Pre-basic (PB)	Progeny of the breeder seed, usually produced by the breeder. This class of seed is commonly used where crops have low multiplication rates and where large quantities of seed are required	85	99
Basic (B)	Progeny of the breeder seed or pre-basic seed. Produced under supervision of the breeder or their agent. Under the control of seed quality agency		
Certified seed (CS)	Progeny of the basic seed and produced under contract by selected growers. Strict quality controls. Can be used to produce further generations of certified seed (C1, C2) or to produce a crop	85	98
Certified seed (C1)	As above, but F_1 generation seed		
Certified seed (C2)	As above, but F_2 generation seed		

by the Home-Grown Cereals Authority (HGCA) under the Agricultural and Horticultural Development Board (AHDB), which publishes recommended lists for UK cereals and oilseed crops and can be found on their website (https://cereals.ahdb.org.uk/varieties.aspx). During seed variety development, any new variety must pass through two key tests, namely Distinctness, Uniformity, and Stability (DUS), and Value for Cultivation and Use (VCU).

The DUS test is a descriptive assessment of variety characteristics using morphological characteristics that are diagnostic of the new variety. The DUS test also assesses the uniformity and stability of the new variety. Any new variety is compared with existing varieties to establish the distinctive characteristics of the new variety, and subsequently a variety description is prepared. The DUS test usually runs for 2 years, and is a useful tool for the production and protection of new varieties.

The VCU test focuses on the agronomic features of the new variety (this information is useful both for the grower and for the end user), such as disease resistance, and looks at features such as grain weight and protein content. The VCU test normally runs for 3 years, and these tests are conducted at multiple locations throughout the country to establish the response of the new variety to regional variations in climate.

Thus only cereal seeds that have been officially certified via crop inspection can be traded in the UK and other European countries. Farmers can save their own seed, but are strongly advised to follow the certification procedures to ensure the purity and health of their stock and future grain harvest. In the UK the certifying bodies are the Department for the Environment, Food and Rural Affairs (Defra) in the case of England and Wales, Science and Advice for Scottish Agriculture (SASA) in the case of Scotland, and the Department of Agriculture and Rural Development (DARD) in the case of Northern Ireland. The certification protocols need to be rigorous to ensure the quality and health of the harvest, as this is obviously directly related to the health of the sown seed.

Seed treatments (Table 10.3) are then generally applied to cereal seeds to ensure strong crop establishment. However, organic cereals are not treated with these seed dressings, so seed health and quality is a fundamental aspect of disease control on organic cereal farms. Failure to implement rigorous seed biosecurity will ultimately result in significant crop losses from germination and establishment right through to harvest.

Table 10.3 Examples of seed dressings used in arable crops, including trade name, active ingredients and their chemical structure, crops protected, and disease and agronomic features of the seed dressing

Trade Name	Active Ingredient	Type	Crop	Target Disease
Galmano®	Fluquinconazole	Triazole fungicide	Winter wheat and barley	Seed-borne and soilborne bunt, early foliar disease, particularly rust, and provides protection against take-all in second wheats
Agrichem HY-PRO Duet	Prochloraz and thiram	Imidazole fungicide and tetramethylthiuram disulfide, respectively	Main crop in UK is winter oilseed rape	Damping off disease and *Phoma*
Redigo Deter	Clothianidin and prothioconazole	Clothianidin is the insecticide component, consisting of nitroguanidine, a subgroup of the nicotinoids. Prothioconazole is based on a triazolinthione structure	Wheat, barley, oats, rye, triticale, and durum wheat	Early control of seed- and soilborne disease—includes bunt (*Tilletia tritici* and *T. caries*, respectively), *Microdochium* seedling blight, *Fusarium* seedling blight (*Fusarium graminearum* and *F. culmorum*), loose smut, and take-all. Early control of foliar disease—includes *Septoria tritici*, yellow and brown rust. Early pest control—includes aphids (vectors of barley yellow dwarf virus), wireworm, slugs, and wheat bulb fly
Jokey®	Fluquinconazole and prochloraz	Triazole and imidazole fungicide, respectively	Wheat	Seed and soilborne disease such as *Fusarium* and bunt. Protection from early-season brown and yellow rusts, delays early development of *Septoria tritici*, and provides control of take-all. In addition, improves standing power of the crop by increasing root length. Agricultural Development Advisory Service (ADAS, 2009)

Potato

Continuing the theme of propagule certification, another significant staple crop is the potato (*Solanum tuberosum*), which is of course mainly grown from vegetative tissue (the tuber), commonly called the seed potato (**Figure 10.2**). One of the major problems with producing a crop from vegetative seed potatoes is that these propagules can carry significant amounts of disease inoculum. Potatoes are susceptible to many soilborne diseases (**Table 10.4**), and can also carry disease inoculum that has been transmitted from the parent plant. The purpose of seed potato classification is to either reduce or completely remove sources of carry-over inoculum (Table 10.4). In order to ensure the security of the potato crop, it is essential that seed potatoes are certified disease-free or meet minimal certification standards. Scotland is a world leader in producing certified seed potatoes, and is recognized by the European Union as a Community Grade region, which applies stricter health standards than other regions in the EU. The production of seed potatoes in Scotland and indeed across the UK is controlled by the Seed Potato Classification Scheme (SPCS), which in England and Wales is administered by the Animal and Plant Health Agency (APHA) on behalf of Defra, and in Scotland is controlled by SASA.

The seed production model in Scotland is based on the following sound phytosanitation principles whereby seed potato production can only occur on land that is free from previous potato crops and associated pests and disease:

- Land used for basic seed potatoes must not have been used for potato production for 5 years, and for pre-basic seed potatoes the land must have been free from potato production for 7 years.

- Land must be free from potato cyst nematodes such as *Globodera* species, and pest-free status must be confirmed by taking soil samples before planting out the seed potatoes.

- Land must also have had no occurrence of wart disease (*Synchytrium endobioticum*), a disease that has not been observed in Scotland for over 30 years.

Another strict production criterion in Scotland is the meticulous application of generation rules. Scotland only produces and markets pre-basic and basic seed potatoes, and will only accept into the region seed potatoes that have been grown and classified as Community Grade.

The initial production of seed potatoes in Scotland starts in the laboratory, using micro-propagation protocols, and is known as nuclear stock, from which all subsequent seed stock is produced (Table 10.5). These nuclear stocks undergo stringent testing to ensure that they are free from disease. Once they have been classified as disease-free, further multiplication is carried out in officially approved commercial micro-propagation facilities to produce disease-free mini-tubers known as pre-basic TC, or tissue-cultured pre-basic stock. Pre-basic TC seed stock are selected and grown from tissue-cultured micro-plants and raised in a protected environment (a greenhouse or polytunnel) and in pathogen-free media such as a hydroponic system or sterilized peat-based compost. Following on from the pre-basic TC, the stock can be weaned into the field for classification as pre-basic, and may be grown for up to four generations (PB 1, 2, 3, and 4). During this production process, stocks are checked throughout the growing season and must be classified as 100% pure and true to type (reversion of type can occur in vegetative propagated stock). They must also be completely free of the following diseases:

- Tobacco veinal necrosis strain of potato virus Y

- Severe and mild mosaic virus infections

- Leafroll viruses

- Blackleg (*Pectobacterium atrosepticum* and associated species)

- Witches' broom (*Phytoplasma* species, transmitted by leafhoppers).

Pre-basic stock is not traded outside of Scotland, as it forms the basis of the seed production system, providing a supply of certified healthy material for all stocks.

There are three more stages to the SPCS that operates in Scotland, namely Super Elite, Elite, and Class A (Table 10.5), and each class has minimum standards for key pathogenic organisms associated with the production system (Table 10.6). The application of the SPCS has ensured a continuous supply of high-quality, disease-free seed potatoes that are traded across the EU and further afield.

Soft Fruit and Top Fruit

Similar seed and propagule certification procedures are applied to the fruit production sector, and because fruit crops are perennial, parent stocks can build up a high degree of disease inoculum over time. Therefore any new variety that is bulked up vegetatively from parent stock must undergo a rigorous selection procedure before it is released to the wider growing community. For example, strawberries (an example of soft fruit) can be propagated from runners (Figure 10.2b), and cherries (an example of top fruit) are grafted onto rootstocks (Figure 10.2c). In both cases the donor or parent stock must be certified free from key pathogens, and in addition the daughter (the runner tip in strawberries or the scion in the cherry crop) must also be certified free from disease. The certification and classification scheme for fruit crops is complex and is beyond the scope of this chapter. However, it is of value

Figure 10.2 Vegetative propagation material. (a) Seed potato; (b) strawberry runners; (c) rootstock graft.

Table 10.4 Examples of soilborne inoculum on *Solanum tuberosum*.

Taxonomic Group	Scientific Name	Common Name	Physiological Damage	Quarantine	Source of Inoculum
Fungi (chytrid)	*Synchytrium endobioticum*	Wart disease	Warts grow out from the eyes of tubers, which turn black once removed from the soil	Yes	Overwintering sporangia, which release motile zoospores. Sporangia can persist in the soil for up to 30 years
Nematodes	*Meloidogyne chitwoodi* and *M. fallax*	Root-knot nematode	Causes root-knots and skin blemishes on the tubers. The latter reduce market quality		Potatoes tubers are infested by overwintering populations, which multiply rapidly, producing second and third generations in a cropping season. Root-knot nematode populations expand in a log-linear relationship
	Globodera rostochiensis and *G. pallida*	Potato cyst nematode	Reduces vigor of crop by causing root damage as maturing nematodes induce root cysts		Cysts on roots, which contain many hundreds of eggs. On hatching, the young nematodes infect new crop. Cysts can persist in the soil for up to 30 years, and the eggs inside the cyst remain viable for the same length of time
Bacteria (Gram positive)	*Clavibacter michiganensis* subsp. *sepedonicus*	Ring rot	Rots vascular tissue, causing a discolored or glassy ring around the tuber		Overwinters in volunteer tubers and on plant debris left on cultivation and harvest equipment
Bacteria (Gram negative)	*Ralstonia solanacearum*	Brown rot	Infects the vascular tissue, causing wilting and bacterial ooze. Tubers rot and eyes become distorted	No	Motile soilborne bacteria, also in water courses and irrigation water. Potato is an irrigated crop
Viroid (circular RNA)	Potato spindle tuber viroid		Difficult to diagnose; can look like nutrient deficiency or environmental damage. Foliage is stunted and tubers become elongated and rounded at ends		Transmission is via infected pollen, plant material, and through open wounds, or debris on equipment
Oomycetes	*Phytophthora infestans*	Potato blight	Rapid spreading of watery rot of leaves. Tubers show reddish-brown decay under the skin		Oospores and hyphae survive in tubers and can reinfect new crops or volunteer plants sprouting and survive several rotations to be a viable source of inoculum for successive crops
Bacteria (Gram negative)	*Pectobacterium atrosepticum* is most common (older synonym *Erwinia carotovora* subsp. *astroseptica*), but also could include *Dickeya dadantii*	Blackleg	Two phases of disease development: (1) on young plants soon after emergence; the plants become stunted, and foliage is chlorotic with a stiff and upright habit. Below ground the stem darkens to black, and decays. Infected pith is very susceptible to decay that may extend upwards to above-ground stems; (2) on mature plants, later in the growing season, blackleg appears as black discoloration of previously healthy stems. This is followed by rapid wilting and chlorotic leaves	No	Infected tubers and via certain insect vectors, such as the fruit fly (*Drosophila melanogaster*), and in the field on infected stems, bacteria can survive in the vascular tissue

Causal organism	Scientific name	Disease	Symptoms		Source of inoculum
Fungi	*Phoma exigua* var. *foveata*	Gangrene	A disease of the tuber. Infected tubers develop a thumbprint-like lesion on the surface of the tuber that has defined edges. Decay sets in and the rot associated with the lesions starts as a pinkish water-soaked lesion that eventually turns purple-black	No	Key source of inoculum is from the black dot-like pycnidia on the surface of the haulm. The pycnidia develop vegetative conidia that can be washed from the haulm into the soil, where they infect tubers. Non-certified seed potatoes may be infected, and when planted will develop into diseased plants that continue the cycle of infection. Volunteer potatoes are also key sources of inoculum
	Fusarium species, commonly *Fusarium solani* var. *coeruleum*	Dry rot	A disease of the tuber. Dark brown lesions occur on the surface of the tuber, and these lesions eventually shrink. The flesh below develops a dark coloration and cavities will eventually occur in the tuber		Common soil-dwelling fungi which infect damaged tubers and hence seed potatoes. Disease can persist for up to 9 years following a potato crop
	Botrytis cinerea	Wet rot	Very similar tuber damage to dry rot, but the haulm is also infected, which can reduce leaf tissue and interfere with vascular systems, hence reducing translocation		Infection occurs on the haulm and is worse in irrigated crops. The fungus overwinters as a sclerotium, and in the spring produces conidia that start a new cycle of infection. Tubers become infected by the conidia, and hence non-certified seed tubers can transmit the disease
Fungi	*Polyscytalum pustulans*	Skin spot	Small brown lesions on the surface of tubers. Severe infections will damage eyes and root tissue		Infected seed potatoes are the key source of inoculum. Can survive on plant debris
Fungi	*Rhizoctonia solani*	Black scurf	Development of hard black patches on the surface of tubers, to which soil adheres. Sprouting from infected tubers can be ring-girdled, and shoots will die off, leading to the proliferation of numerous weaker sprouting shoots		A very common soil-dwelling fungus. Can overwinter saprophytically on plant debris as a sclerotium and produce new spores in the spring. Infected seed tubers can act as a source of inoculum
Filamentous bacteria belonging to the Actinomycetes	*Streptomyces scabies*	Common scab	The tuber surface develops numerous lesions and erupted pits. These skin blemishes reduce the market value of the crop	No	A very common soil-dwelling organism in all potato-growing regions. Seed potatoes can act as a source of inoculum
Plasmodiophorales	*Spongospora subterranea*	Powdery scab	Similar symptoms to common scab, but lesions on the tubers are more rounded in outline		The lesions on the tubers are composed of spore balls that are eventually released into the soil. These spores then develop into motile spores that infect root hairs. Infected seed tubers act as a source of inoculum. Spores may survive for 10 years or more in infected soils

Table 10.5 Seed classification scheme for seed potato certification in the UK.

Seed Classification	Production System	Sources of Initial Stock	Number of Generations
Nuclear stock	Micro-propagation (government laboratories at SASA)	UK and EU potato breeders, non-EU potato breeders following quarantine	N/A
Pre-basic TC	Micro-propagation (in approved commercial laboratories)	From nuclear stock	N/A
Pre-basic (PB 1–4)	In approved and disease-/pest-free field locations	Pre-basic TC	4
Super Elite (SE 1–3)		Pre-basic (PB 1–4)	3
Elite (E 1–3)		Pre-basic (PB 1–4) and Super Elite (SE 1–3)	
Class A		Downgraded from second year of PB, SE, or E	N/A

to explore some of the basic principles of the fruit crop certification scheme. This can be achieved by using summary tables (Table 10.7 and Table 10.8), which highlight the key principles of fruit certification, including the grade of certified stock, key pathogens and associated tolerances, and isolation distances for certified stocks used by plant breeders.

Strawberry and apple crops each have their own certification grades relating to the production system, and these are generally arranged in descending order of purity as shown below.

Strawberries

- Foundation—any plant material that meets the requirements of nuclear stock (Table 10.7).

- Super Elite—any variety that was certified at Foundation grade in the previous year.

- Elite—any variety that was certified at Foundation or Super Elite grade in the previous year.

- "A"—any variety that was certified at Foundation, Super Elite, or Elite grade in the previous year.

Table 10.6 Acceptable disease tolerances for UK seed potatoes.

Seed Classification	Purity and True to Type (%)	Pathogens	Tolerances (%)
Nuclear stock	100	All pathogens	0
Pre-basic TC		All pathogens	0
Pre-basic (PB 1–4)		All pathogens	0
Super Elite (SE 1–3)	99.95	Severe mosaic virus	0
		Leafroll virus	0.01
		Mild mosaic virus	0.05
		Blackleg (*Pectobacterium* species)	0.25
		Blackleg (*Dickeya* species)	0
Elite (E 1–3)	99.95	Severe mosaic virus and/or leafroll virus	0.1
		Total mosaic and leafroll virus	0.5
		Blackleg (*Pectobacterium* species)	0.5
		Blackleg (*Dickeya* species)	0
Class A	99.90	Severe mosaic virus and/or leafroll virus	0.4
		Total mosaic and leafroll virus	0.8
		Blackleg (*Pectobacterium* species)	1.0
		Blackleg (*Dickeya* species)	0

- Approved Health—any plant material produced vegetatively or by seed but not on a Pedigree variety list (which is constantly changing over time). All vegetative stock in the Approved Health category must be tested for freedom from disease (Table 10.7) before being planted out into production beds, unless the parent plants tested disease free in the previous year.

Table 10.7 Acceptable disease tolerances for vegetative propagation material used in the UK soft fruit industry.

Taxonomic Group	Scientific Name	Common Name	Notifiable	Plants can be Rogued within the Production System	*Foundation and Super Elite			Elite		A and AH	
					Crop Inspection Sequence						
					First	Second	Third	First	Second	First	Second
Virus	N/A	Arabis mosaic virus	Yes	No	0%	0%	0%	0%	0%	0%	0%
		Raspberry ringspot virus									
		Tomato black ring virus									
		Strawberry crinkle virus									
		Strawberry latent ringspot virus									
		Strawberry mild yellow edge virus									
		Other virus disease	No		0%	0%	0%	0.2%	0.2%	5.0%	5.0%
Phytoplasma	N/A	Strawberry phytoplasma	No	No	0%	0%	0%	0.2%	0.2%	5.0%	5.0%
Bacteria	*Xanthomonas fragariae*	Angular leaf spot	Yes	No	0%	0%	0%	0%	0%	0%	0%
Oomycota	*Phytophthora fragariae*	Red core	Yes	No	0%	0%	0%	0%	0%	0%	0%
	P. cactorum	Crown rot	No	No	0%	0%	0%	0%	0%	0.5%	0.5%
Fungi	*Verticillium dahliae*	Verticillium wilt	No	Yes	0%	0%	0%	5.0%	5.0%	5.0%	5.0%
	Colletotrichum acutatum	Strawberry blackspot	Yes	No	0%	0%	0%	0%	0%	0%	0%
Nematodes	*Ditylenchus dipsaci*	Stem eelworm	No	No	0%	0%	0%	0.5%	0.5%	2.0%	2.0%
	Aphelenchoides ritzemabosi	Leaf eelworm			0%	0%	0%	0%	0%	2.0%	2.0%
	A. fragariae				0%	0%	0%	0%	0%	2.0%	2.0%
Mites	*Phytonemus pallidus fragariae*	Strawberry tarsonemid mite	No	Yes	0%						
Other pests and diseases, including mildew, aphids, capsid bugs, two-spotted spider mite, and caterpillars			Substantial freedom from infestation								

*Categories of crop grading scheme – Foundation grade, Super Elite grade, A grade, AH (approved health) grade – the latter = inspection for health and vigor only.

Table 10.8 Acceptable disease tolerances for vegetative propagation material used in the top fruit industry in the UK.

Plant Characteristics	Taxonomic Group	Scientific Name	Common Name	Notifiable	All Grades		
					Mother Trees	Rootstocks	Complete Trees
Purity	N/A				0%		
True to type (not N grade)	N/A				0%		
	Virus	N/A	Plum pox virus	Yes	0%		
			Other viruses	No			
	Bacteria (Gram negative)	*Erwinia amylovora*	Fireblight	Yes	0%		
	Fungi	*Chondrostereum purpureum*	Silver leaf	No	If more than 10% of branches are infected, tree will need to be grubbed	Any plant showing symptoms will not be certified	
	Other pests and diseases, including mildew, apple scab, aphid, and spider mite				Substantial freedom from infestation		

Apples

Certification of mother trees (note that N in parentheses indicates a nominal grade):

- Trees budded with a scion from an approved source.

- Super Elite (N)—any Super Elite not listed in the pedigree list of main varieties (https://www.gov.uk/guidance/fruit-propagation-certification-scheme). This includes ornamental fruit trees.

- Elite (N) as outlined above, but in descending order of rigor.

For new mother trees:

- Super Elite and Super Elite (N) which is bud wood from an approved source, including re-tested Super Elite trees, as well as rootstocks certified at Foundation grade.

- Elite and Elite (N) which is healthy bud wood from an approved source and rootstocks certified at Super Elite and Elite stage. Maiden trees must meet all standards and requirements.

For established mother trees:

- Super Elite and Super Elite (N) which is stock that was certified Super Elite in the previous year.

- Elite and Elite (N) which is stock that was certified Elite or Elite (N) in the previous year.

To increase stock:

- Maiden trees produced on the grower's land or nursery from certified bud wood of mother trees or from certified and approved external sources and Foundation grade rootstocks.

- Elite and Elite (N) maiden trees produced from certified bud wood from the grower's own stock and Super Elite and Elite grade rootstocks.

The grading system for apple certification outlined above is applied to the development of rootstocks and scions within the plant production system. Furthermore, stock can be certified from imported sources if the imported material meets the minimal criteria for each of the grades listed above (Table 10.8). In addition to the above, there are restrictions on how old trees can be in relation to their given grade. For example, mother trees in either the Super Elite or Super Elite (N) grades must be no more than 7 years old, and the age of trees in the Elite and Elite (N) grades must not exceed 20 years.

A final comment on fruit certification relates to planting distances in fields and/or protected cover production, such as use of gauze covering. Crops of all fruits have specified planting distances for each category or grade, and these are applied across the categorical scheme and can range from as little as 3 m apart between Foundation vs. Foundation grade in strawberries to 1500 m apart between the Foundation category for fruit that is going for passport or fruiting certification. The above discussion and associated tables on fruit certification schemes provide a summary of the key concepts (a full description of the scheme for each type of fruit grown in the UK is available at www.gov.uk/guidance/plant-health-propagation-scheme).

10.3 Breeding for Disease Resistance

Following on from biosecurity and seed certification, the next tool in the battle against plant pathogens is plant breeding. This is the linchpin of disease management for organic growers, but it is also a vital tool for conventional growers, as disease-resistant cultivars help to suppress the spread of disease across a growing region. A good example is the use of the cultivar diversification scheme adopted by wheat growers in the USA, in which different cultivars of wheat, with varying degrees of disease resistance, are grown within and across a state and across neighboring states in order to reduce the ravages of wheat rusts such as stem rust (*Puccinia graminis* f. sp. *tritici*). Stem rust first appears in the southern states around early May. Wind-borne spores then spread north, and by June 24 rust is infecting wheat in the northern states and the border with Canada. Plant breeding is a time-consuming process, and it may take 10–12 years to develop a new resistant cultivar with commercial value using traditional plant breeding methods (outlined below). However, modern molecular techniques such as quality trait loci mapping can reduce the time needed to develop a new cultivar.

Before a breeder can begin to develop a new disease-resistant cultivar or plant line, they need to take into account the following considerations to ensure that their new cultivar has commercial value.

- **Commercial crops need to be uniform in growth and development.** This is essential, as a non-uniform crop is difficult to manage and presents significant logistical problems at harvest if the crop has matured at significantly different rates.

- **Crops must be high yielding.** Clearly, in view of the competitive commercial markets within which growers have to trade, it is necessary to continue to improve crop yields in order to maintain food security.

- **Crops must make efficient use of managed resources.** A new cultivar must be able to access and redistribute managed resources, such as applications of crop nutrients, to ensure uniform growth and development of the crop and uniform quality of the harvestable component.

- **Crops must have a good degree of disease and pest resistance.** It would be self-defeating to release a new cultivar that has improved disease resistance to the rusts, for example, but only weak or moderate resistance to the array of common diseases within the target cropping system. In this case crop health would be significantly compromised by these remaining pathogens, and in worst-case scenarios complete crop failure could occur.

- **Quality of the harvestable component.** All crop plants have specific criteria with regard to the quality of the harvestable component. These criteria can be as diverse as protein content, taste, and cookability (which is particularly relevant in relation to potatoes).

However, breeding for disease resistance is conceptually different to breeding for abiotic stress and other important commercial traits, such as yield, taste, and size of the harvestable component, because breeding for disease and pest resistance needs to take into account the dynamic processes that are involved in the host–pathogen interaction. This interaction is dominated by the ways in which the host responds to the pathogen, and these responses are mediated by genes known as resistance genes (*R* genes), which may be either race specific or race non-specific. Race-specific genes are associated with a type of resistance known as vertical resistance, and race-non-specific genes are correlated with horizontal resistance.

Vertical resistance refers to a resistance process that is controlled by one or a few major *R* genes and which brings about an immediate and major resistance reaction in the host, such as the hypersensitive response (see Chapter 4). Vertical resistance is also known as **race**- or **pathotype**-specific resistance, and is not particularly durable once it is released into a commercial system. Many of the common cereal diseases, such as the rusts, powdery mildews, smuts, and bunts, appear to be race specific and follow the gene-for-gene concept outlined by Flor (1956). The resistance reaction in a host that expresses vertical resistance is often quantifiable in that when a particular race of a pathogen infects a range of cultivars, some cultivars will display disease symptoms while other cultivars exhibit no symptoms. It is this differential expression of disease symptoms that forms the basis of many crop improvement programs and the development of new disease-resistant cultivars, and globally a good model is the development of new disease-resistant cereals. However, as stated above, vertical resistance has limited durability. For example, race-specific resistance to rusts generally lasts for about 5 years, while race-specific resistance to smuts and bunts lasts a little longer. Breeding for vertical resistance has been successfully applied to most annual crops, but has no application in the breeding of perennial crops.

Horizontal resistance (also termed partial resistance or field resistance) refers to a resistance reaction that is controlled by numerous minor genes, so is polygenetic, and presents a race-non-specific scenario between host and pathogen. In horizontal resistance, infected hosts often display disease symptoms such as rusting, but these symptoms are slower to develop and have been termed "slow rusting." Because horizontal resistance is polygenetic in nature, it is more durable than vertical resistance, and the host response to disease challenge under horizontal resistance is typically minor to intermediate. Achieving new cultivars that express horizontal resistance is the desired trait in any crop system, due to the durability of the race-non-specific nature of this resistance, and breeding programs that target the horizontal resistance reaction are ideal for perennial crops.

Approaches to Plant Breeding

Successful release of a new improved cultivar involves a number of sequential steps that should be followed to ensure the quality and durability of the new cultivar. Breeders are aiming for durability of resistance, which would typically require horizontal resistance. Often, however, it is vertical resistance that dominates in new cultivars, such as the utilization of a race-specific approach to the wheat improvement program.

The first step is to assemble a collection of *R* genes (Research Box 10.1 and Table 10.9), identify their mechanisms of action (that is, race specific or race non-specific), and then maintain the collection in botanical gardens and

RESEARCH BOX 10.1 PLANT DISEASE RESISTANCE GENES

Introduction

Over the last 30 years there have been significant advances in our knowledge of plant disease resistance genes (see Chapters 3, 4, and 10), which have helped to improve our understanding of how plants resist disease, and have facilitated the development of new commercial cultivars by plant breeders. The gene-for-gene theory (see Chapters 3 and 4) proposed by Flor (1956) postulates that for every pathogenic challenge and resistance response observed in plants there must be a complementary set of interacting genes (**alleles**) in the host–parasite partnership. These genes were assigned as either resistance genes (*R* genes) in the host, or avirulence (*Avr* genes) in the parasite. For a resistance reaction to occur—that is, for what is confusingly termed an incompatibility reaction in the host (Figure 4.1)—there must be a set of *Avr* genes in the parasite. The natural application of the gene-for-gene theory is to develop new disease-resistant cultivars, and this has become the main strategy used by commercial growers to combat disease.

R Genes and R-Gene Products Explained

R genes are a series of genes where the alleles have been characterized across known plant chromosomes. Several types of *R* genes (Table 10.12) have been isolated from a range of crop species, and their protein products have been fully described or a putative function has been proposed for them (Table 10.12). The main principle is that *R* genes enable the plant to (1) detect the *Avr*-gene products from the parasite, (2) initiate signal transduction to activate defenses (Figure 4.1), and (3) have the capacity to evolve new *R* genes and their associated products rapidly following the evolution of new *Avr* genes in the parasite. A full molecular description of *R*-gene activity is beyond the scope of this book, but it is possible to describe the function of two of the more abundant *R* genes documented in the literature, namely the *RPM1* gene from *Arabidopsis thaliana*, which confers resistance to *Pseudomonas syringae* that carries the corresponding *avrRpm1* gene, and the *N* gene from tobacco (*Nicotiana tabacum*), which confers resistance to most strains of tobacco mosaic virus (TMV) which has the corresponding **TMV-derived ligand** (the elicitor of pathogenesis in TMV infections, analogous to *Avr* genes).

The *RPM1* gene is a class of *R* genes that are described by a series of conserved motifs. In the case of the *RPM1* gene

the nomenclature of this gene is known as an NBS-LRR class, where NBS denotes nucleotide-binding site and LRR denotes leucine-rich repeats. The described structure of the *RPM1*-gene product begins at the amino terminal with a leucine zipper region, followed by the NBS region, then an internal hydrophobic domain, and finally at the carboxy terminal a leucine-rich repeat (Figure 1). The structure of the *RPM1* gene is not unique in the wild, and homology has been observed in the *A. thaliana RPS2* gene, another *R* gene that confers resistance to *P. syringae*, and elements of the *RPM1*-gene structure are not uncommon in the *N*-gene of *N. tabacum*.

- The *N*-gene product in tobacco consists of a TIR domain at the amino terminal, an internal hydrophobic region, then an NBS region, and at the carboxy terminal an LRR (Figure 1). The TIR element relates to the homology observed in the cytoplasmic domains of the *Drosophila* and human Toll protein and the interleukin-1 receptor (IL-1R) in mammals. The standard nomenclature for TIR is thus given as TIR = Toll, IL1R. The *N* gene may also be a truncated protein N^{tr}, which has the same structure as the *N* gene but codes for a protein product consisting of 652 amino acids compared with the 1144-amino-acid protein of the *N* gene. The presence of the NBS and LRR in the *N*-gene products places the *N* gene within the NBS-LRR class of plant *R* genes.

The above examples of *R* genes have been classified into the NBS-LRR family (nuclear-binding site–leucine-rich repeats), but there is a large range of *R*-gene families that are all described by their protein domains (Table 10.12). The vast majority of these *R* genes fall into the categories of NBS-LRR, eLRR, or LRR-kinase superfamilies, where the functional domains of the NBS and the core of the LRR elements are highly conserved across a large array of plant species. Indeed the NBS-LRR, eLRR, and LRR-kinase families of *R*-gene products are ubiquitous in the plant kingdom. Although the LRR domain of these *R*-gene products is known to be highly conserved, it has been demonstrated that there is considerable variation in the peripheral regions of the LRR domain. It has been postulated that the variation in the amino acid sequence in the peripheral regions of the LRR domain may account for the rapid response of these major gene products

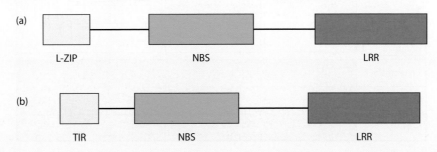

Figure 1 Schematic representation of (a) the *RPM1* gene from *Arabidopsis thaliana* and (b) the *N* gene from *Nicotiana tabacum*. (Adapted from the following: Boyes et al., 1998; Erickson et al., 1999; Dinesh-Kumar et al., 2000.)

RESEARCH BOX 10.1 PLANT DISEASE RESISTANCE GENES

to the evolution of resistance to plant defense mechanisms, observed in the pathogen population, when these pathogens have been challenged to the strong selection pressure of single (vertical resistance) gene disease resistance. The location of these *R*-gene families within plant tissue is variable, and it has been reported that many are cytoplasmic, transmembrane, and have functional domains in the extracellular spaces.

The complexity of *R* genes and their associated protein products is significant, but the salient point is that these molecular switches enable plants to defend themselves from pathogenic challenge by a range of known reactions that are typified by symptoms expressed in plants after they have been infected by pathogens. The mechanism of action of these *R*-gene products is the subject of ongoing research, but a few models of how they function have been postulated (Figure 1). These putative mechanisms are based on the concept of signal transduction and the downstream up-regulation of plant defense mechanisms (**Figure 2**). The key point is that these *R* genes can be traced through well-designed plant breeding programs and are indeed the subject of ongoing development of new disease-resistant cultivars and hybrids across all of the plants used by humans.

Incorporating *R* Genes into New Cultivars

Using the *Lr* genes (leaf rust genes) of wheat as a model system, plant breeders have been able to document the incorporation of numerous *Lr* genes into commercial cultivars of wheat over many decades of crop improvement programs. A total of 25 *Lr* genes were originally isolated directly from **hexaploid** wheats (Kolmer, 1996). A number of the *Lr* genes have been derived from wheats with a lower ploidy, relatives of hexaploid wheats, within the Triticeae **tribe** from the family Poaceae. Certain *Lr genes*, such as *Lr35* and *Lr36*, were isolated from *Triticum speltoides*, and another cluster of

Lr genes (*Lr39*, *Lr40*, *Lr41*, *Lr42*, and *Lr43*) were isolated from *T. tauschii*. The breeding system used to consolidate the *Lr* genes often utilizes the backcross methods (for example, the *Lr35* gene was transferred by backcrossing the **amphiploid** of *T. speltoides* × *T. monococcum* to the wheat cultivar known as Marquis). Many of these *Lr* genes were consolidated into modern cultivars by the backcross method. Evaluating the success of these gene transfers into wheat cultivars using traditional plant breeding strategies requires the breeder to monitor the infection type using a scoring scheme that starts with a simple scale, say from 0 (no visible uredinia or flecking symptoms) to 4 (large uredinia). These scoring schemes can be quite complex and have numerous subscripts that correspond to distinctive disease symptoms (Kolmer, 1996). Finally, the expression of disease symptoms is also related to plant age, dominance relationships of genes, temperature effects, and suppressor genes.

Modern recombinant DNA techniques such as qualitative trait loci mapping can enhance the isolation and transfer of *R* genes into new plant lines, and it is now possible using vectors such as *Agrobacterium tumefaciens* to transfer *R* genes from a range of diverse species into important commercial crops that may lack key *R* genes. For example, Yang et al. (2013) transferred *R* genes from maize (*Zea mays*), sorghum (*Sorghum bicolor*), and brachypodium (*Brachypodium distachyon*) into rice (*Oryza sativa*) susceptible to the rice blast pathogen (*Magnaporthe oryzae*). A total of 16 large *R*-gene families were transferred into rice, and all of these cloned genes had the NBS-LRR domains. Following successful transfer into rice, these transgenic plants expressed a range of resistance reactions to the rice blast pathogen, and the researchers were able to demonstrate that 25% of their cloned genes were resistant to at least one of the 12 races of rice blast.

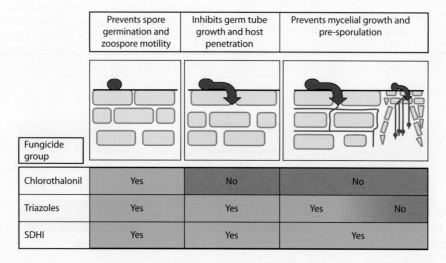

	Prevents spore germination and zoospore motility	Inhibits germ tube growth and host penetration	Prevents mycelial growth and pre-sporulation	
Fungicide group				
Chlorothalonil	Yes	No	No	
Triazoles	Yes	Yes	Yes	No
SDHI	Yes	Yes	Yes	

Figure 2 Schematic representation of disease control by key fungicides on fungal phases of the pathogen life cycle. SDHI, succinate dehydrogenase inhibitors. A color version of this figure can be found in the color plate section at the end of the book.

Table 10.9 Examples of *R*-gene domain abbreviations, with their full nomenclature and putative function.

R-gene Abbreviation	Full Description of *R*-Gene Domains	Function of *R*-Protein Domains
CC-NBS-LRR	Coiled-coil–nucleotide-binding site–leucine-rich repeat	CC: Typically consists of two or more α-helices wrapped around each other to form a superhelical twist with both hydrophobic and charged polar residues. CC domains are key elements in protein–protein interactions and form part of the R–Avr specific recognition processes. CC domains are present in both monocots and dicots
LZ-NBS-LRR	Leucine zipper–nucleotide-binding site–leucine-rich repeat	NBS: The NBS domain is found in many ATP- and GTP-binding proteins, and suggests possible activation of phosphate-binding and transfer-type reactions. NBS plays an important role in plant signaling. In tomato, two R proteins with NBS domains confer resistance to *Fusarium oxysporum*
NBS-LRR	Nucleotide-binding site–leucine-rich repeat	TIR: This domain was originally characterized due to its homology with intracellular regions of the mammalian IL-1 receptor and the *Drosophila* protein Toll. The Toll-like proteins and IL-1R signaling pathways are key mediators of the innate immune response in both *Drosophila* and mammalian taxa infected with bacteria and fungi. It is believed that the TIR domain in plants initiates conserved signal transduction pathways in response to pathogens
TIR-NBS-LRR	Toll interleukin-1 receptor–nucleotide-binding site–leucine-rich repeat	LRR: R–Avr protein recognition. It is currently known that LRR domains contain at least seven distinct subfamilies, which provide a range of structural arrays that enable the LRR domains to form protein–protein interactions. Experimental evidence demonstrates that LRR domains are involved with effector proteins and are key to the specificity of resistance. TIR domains are common in dicots but rare in monocots, and no TIR domains have been identified from the rice genome

nurseries. Sources of *R* genes are varied, ranging from extant commercial cultivars, landraces, related wild species, and forebears to obsolete commercial cultivars (as these obsolete cultivars will already contain a range of desirable traits related to commercial viability, as discussed above) which will have fewer undesirable traits compared with wild species. Once the germplasm has been established, the next phase involves the use of one of the following breeding strategies:

- Self-pollinated species, leading to pure inbred lines

- Cross-pollinated species, leading to a broad and diverse genetic base

- Hybrid breeding program, leading to superior vigor in the offspring.

Each of the breeding strategies listed above involves a range of approaches for achieving a successful new cultivar.

Breeding Self-Pollinated Species

The advantage of breeding self-pollinated species is that the breeder can quickly develop a homozygous and homogenous plant line, which is an important commercial parameter in crops such as the cereals, where uniform crop development is desirable. The most common approach to breeding self-pollinated species is the use of the mass selection procedure, which is one of the oldest breeding systems in use, dating back to the origins of plant domestication. Mass selection is based on the concept of selecting from a biologically variable population those plants that exhibit desirable traits, and discarding those plants that exhibit undesirable traits. The aim of mass selection is to improve the frequency of desirable genes relative to the undesirable genes, and it relies on the undesirable plants being rogued out and discarded from the population. This is achieved by growing on the selected plants in progeny rows in year 2 of the selection process. The selection process can be repeated multiple times (this is known as recurrent mass selection). Selection of desirable plants by the mass selection method relies on the expression of the phenotype, and thus is effective where the trait of interest has a high degree of heritability.

A schematic representation of the general steps in mass selection is shown in **Figure 10.3**.

Thus mass selection develops homozygous lines and breeds out heterozygous alleles, but this process is not 100% pure, as the new cultivar could still retain significant variability for measurable traits, and a non-uniform growing environment exacerbates this phenomenon. Mass selection can be applied to cross-pollinating species as well as to self-pollinating species. Using species that self clearly aids the development of homozygous lines, whereas applying mass selection to cross-pollinating species will always lead to heterozygous lines. There are a number of advantages and disadvantages to the mass selection method (**Table 10.10**). Mass selection can then be further modified and developed to produce a range of other breeding programs, such as pure-line selection and the pedigree selection program.

Pure-line selection or breeding occurs when a cultivar is identical to the parent, with a coefficient of parentage of 0.87 meaning that the new cultivar has identical alleles to the parent at all loci. This is of course not easily achieved, as mutation rates will introduce new alleles over time. Pure-line selection is based on repeated selfing following the initial selection from a mixture of homozygous lines. Pure-line selection is a powerful tool that has been applied to both the agricultural and horticultural sectors, where

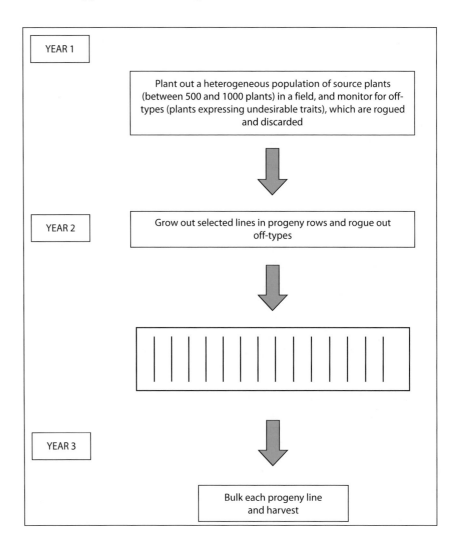

Figure 10.3 Schematic representation of mass selection plant breeding program. (After Acquaah, 2012.)

Table 10.10 Advantages and disadvantages of different plant breeding programs.

Advantages	Disadvantages
Mass selection	
A rapid process that enables breeders to handle large populations in one generation cycle	To be effective, traits of interest must have a high degree of heritability
The new cultivar is phenotypically uniform. However, the genotype is variable as it is a mixture of pure lines, and hence the new cultivar is genetically broad based and thus adaptable and stable	As the selection is based on phenotype, it is important that the breeding program is conducted in a uniform environment
The process is inexpensive	Phenotypic uniformity is less than that observed in pure-line programs because mass selection does not incorporate repeated cycles of selfing
	Where genes are dominant, heterozygous lines are indistinguishable from homozygous dominant lines. Without progeny testing, the heterozygote lines will segregate out in the next generation
Pure-line selection	
The base population can be a landrace, and the population size can be large or small depending on the aims of the breeding program	Cultivar purity may be compromised by natural crossing and mutations
Only the best pure line is selected, in contrast to the more variable selection in the mass selection program	The cultivar has a very narrow genetic base and hence is very susceptible to changes in growing environment, such as climate change, and evolution of new pathogens, including new races of common disease and pests
The final cultivar has high uniformity and is ideal for the modern discerning marketplace where crops are graded by their morphological appearance	No new genotype is created; there is merely improvement through the isolation of the key desirable traits, and this method leads to genetic erosion
The method is amenable to breeding in traits that have low heritability, as selection is based on progeny performance	Pure-line selection takes up more time and space than mass selection because of the extra rows required for progeny testing
Pedigree selection	
The requirement to keep detailed records helps breeders to develop a database of hereditary information	Record keeping is onerous and time consuming
Selection uses both the phenotype and the genotype, thus making selection of superior lines from a segregating population more robust	This program requires more space than either mass selection or pure-line selection, and is not suitable for species in which individual plants are difficult to isolate and characterize
A high degree of genetic purity is obtained and thus the plant breeder's rights can be certified	Pedigree selection is a long process, and it can take up to 12 years to develop a new cultivar
	The method is more suitable for qualitative than for quantitative disease resistance, as it is not effective in accumulating numerous minor genes
	Selecting at F_2 for yield is not robust, hence the need for F_3 generations where yield is selected from progeny rows rather than individual plants as in the F_2
Backcross selection	
The number of field testing sites needed is reduced, as the backcross will be adapted to the same area as the donor	The trait of interest should be highly heritable and easy to identify in each generation
If the same parents are used, the same backcrossed cultivar can be recovered	The presence of undesirable linkages can prevent the cultivar from being improved for key agronomic traits of the original donor
It is useful for introgressing specific genes from wide crosses	It is very time consuming to transfer recessive traits
It can be used in both self-pollinated and cross-pollinated breeding programs	

the progeny of pure-line selection form the basis of many important breeding efforts, such as the development of genetic stock for disease resistance, nutritional quality, and cultivar development for the discriminating marketplace. Pure-line selection follows the same initial steps as described for the mass-selection process, but there are additional years (years 3–10) in which the quality of the hereditary trait is assessed and compared with that of the extant commercial cultivar.

Pedigree selection is a method that is applied to both self-pollinating and cross-pollinating species, and it is used to generate new hybrids where reliable and identifiable traits are targeted for improvement. Unlike mass selection and pure-line selection, the initial cross in the pedigree selection program is a hybridization step (Figure 10.4) that is used to introduce genetic variability in the base population. From this initial hybridization step there are two possible outcomes at the first generation (F_1). Either the F_1 plants are the target of commercial release (for example, numerous ornamental flowering plants and garden vegetables), or the F_1 plants are used to grow a plentiful supply of F_2 plants (Figure 10.4) which are then used to select desirable traits. The desirable F_2 plants are grown to harvest, and good record keeping is essential. From the F_2 selected plants the F_3 generation leads to further elimination of undesirable plants as the superior F_3 plants are selected for the F_4 generation, where they are grown in rows from which 3 to 5 plants are selected and planted out into family or progeny lines in the F_5 to F_6 generations. The next generation, F_7, is the beginning of yield trials, where the F_7 is a basic trial from which superior plants progress to replicated and advanced yield and agronomic trials during the F_8 to F_{10} generations, and finally cultivar release around the end of F_{10} (Figure 10.4).

Pedigree selection involves a significant record-keeping burden, as it is essential to track parentage, plant typing for key agronomic characteristics

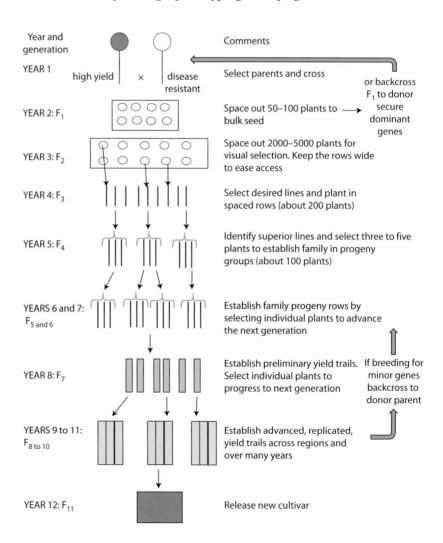

Figure 10.4 Schematic representation of hybrid and backcross breeding program. (After Acquaah, 2012.)

such as yield and harvest quality, growth and development, and key phenotypic characters such as straw length in the cereals, or curd development in crops like the cauliflower. The main advantage of the pedigree program is that selection is based not only on phenotype but also on genotype, making this method of plant breeding an effective tool for selecting superior lines from a segregating population (Table 10.10).

Backcrossing is another procedure used in the self-pollinated program to integrate a new gene or trait into an established cultivar that lacks the desired trait but has all of the other key characteristics for the specific commercial market. Backcrossing is a very useful tool for developing a new cultivar from one that lacks a major and dominant trait, such as disease resistance, which can be easily detected in the new hybrid at the phenotype level. The backcross takes place from the F_1 stage where the new hybrid is backcrossed to the donor or parent of interest for dominant genes, and for recessive genes the backcross takes place again at around the F_8 and F_{10} stages of the breeding program (Figure 10.4).

Breeding from a cross-pollinating population is very complex, and a detailed description of the main methods used is beyond the scope of this chapter (the interested reader is referred to Acquaah, 2012). However, the pedigree-selection program described above (Figure 10.4) is one of the main methods used in cross-pollinating programs, along with recurrent selection. Recurrent selection refers to a method in which systematic selection of individuals expressing desirable traits occurs over several years, where the selected individuals are mated to form a new population that progresses to the next stage in the breeding program. The aim of the recurrent breeding program is to improve the performance of a crop or target plant group by increasing the mean performance of the individuals in that population with regard to the desired trait or suite of desired traits that are being selected for.

The hybrid-breeding program is an important tool in plant breeding as it allows breeders to develop offspring with superior vigor. For example, the development of the modern maize crop (*Zea mays*) has seen significant improvements in crop yield since the 1950s as a result of the hybrid-breeding program. Again a full description can be found in Acquaah (2012).

10.4 Monitoring and Disease Assessment

Another crucial tool that is used in disease management is assessment of the presence and severity of plant disease throughout the crop, garden, orchard, or forest coupe. Accurate identification of a particular pathogen, followed by repeatable assessments of the abundance of the pathogen on a single plant, and then scaling this up across a crop requires years of field experience. Merely identifying a particular species of pathogen can be quite complex, and may require the input of an experienced pathologist or the acquisition of field samples for laboratory analysis, which is frequently the case in the identification of pathogens such as the numerous *Phytophthora* species, as several different *Phytophthora* pathogens may infect a single plant species (Table 5.6). Advances in biotechnology have seen the development of enzyme-linked immunosorbent assay (ELISA) tests, which were initially developed in the 1970s for protein analysis in plant science laboratories, but are now employed as quick and reproducible field kits for growers and agronomists to use in the field. However, after some training and supervised field experience, certain crop pathogens (for example, the mildews, the rusts, *Septoria*, rose black spot, apple scab, and gray mold) are easy to identify. Some of these pathogens cause symptoms within a crop early on. In particular, some of the mildews, rusts, and *Septoria* species can be quite problematic in winter-sown arable crops if conditions during the autumn and winter months are wet and warm. Such conditions favor the activation of resting spores and the development of disease in the lower canopy early in the cropping season. If such a disease threat is not detected and so is left untreated, the grower will most probably lose the entire crop. Therefore it is essential to identify and monitor the incidence and progression of a disease throughout the cropping season, and this critical disease management tool requires the grower or agronomist to assess the level of disease threat within a crop and/or cropping region. To achieve this, it is essential to monitor the spread of a disease across the crop by making assessments of disease thresholds on samples taken across this region. This process requires repeatable estimates of disease thresholds to be made on samples; this simply involves assigning a score on a percentage scale to the degree of disease cover across a range of stems and leaves. In the early stages of a disease this is quite straightforward, but as the disease progresses the infected leaves and stems age, and the number and size of stem and leaf lesions increase, so it becomes increasingly difficult to determine the percentage of disease cover. This presents problems when predicting when to spray, as many crop diseases have quite a low threshold of disease severity before they cause economically significant damage. Indeed, a key point in disease management is the difference between incidence and severity. Incidence is clearly the occurrence or appearance of a disease, which does not necessarily relate to a damaging epidemic. On the other hand, severity is a measure of the abundance of a disease across a crop, and often quite a low level of disease severity can be economically damaging.

Assessing Disease

Yield loss to disease is still quite significant, in the range of 10–16% annually, and the relationship between disease severity and yield loss can be plotted on a graph where disease intensity is represented on the *x*-axis and yield loss is represented on the *y*-axis. However, in order to obtain accurate and replicable information, disease assessment keys must be quick and easy to use, reliable, and reproducible from one field assessment and assessor to another, and between different locations. In order to achieve rapid, reliable, and replicable disease assessment keys, it is essential that these field keys are accurate and precise. Accuracy is a measure of closeness of fit of the disease assessment to the true value observed on a sample, and precision is a measure of the reliability and/or replicability of a disease assessment key between assessors and repeatable use of a given key by one assessor. The development of reliable, accurate, and precise keys has generally followed one of two approaches—either with the relative severity of the disease being categorized on a numerical scale, or with the severity of the disease being expressed as a percentage value.

Presentation of disease assessment keys is typically pictorial (Figure 10.5), but some keys have been developed in the form of a summary table (Table 10.11). The use of disease assessment keys is widespread in the agronomy sector. Furthermore, plant breeders use disease assessment keys to support decision-making processes where breeders are measuring the responses of their new lines to individual physiological races of pathogens such as the rusts. These keys support the selection of new disease-resistant cultivars, as the breeder can measure the response of a given variety to a given race of rusts (Table 10.12).

Sampling Method and Application of Disease Assessment Keys

The literature is replete with discussions of sampling methodologies and how to assess the volume of samples required to make representative estimates of disease threat. The rigorous approach described by research scientists is essential if realistic models of disease spread are to be developed. However, the arguments and sampling methodologies discussed in the literature are

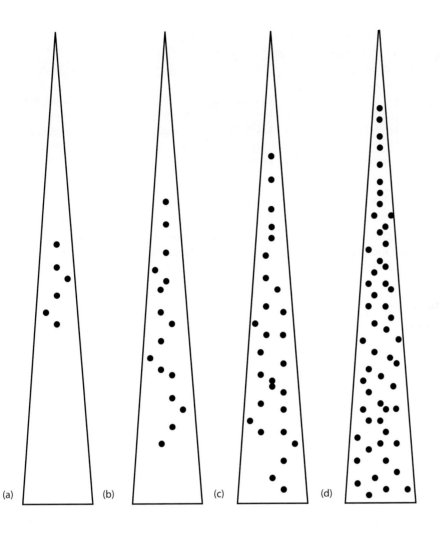

Figure 10.5 Schematic representation of a disease assessment key for leaf rusts of cereals, which can be used for crown rust on oats (*Puccinia coronata*) and for leaf rusts on wheat (*P. triticina*) and barley (*P. recondita*). The percentage leaf area affected is relative to the density of rust pustules and the remaining green leaf area: (a) 1%; (b) 5%; (c) 15%; (d) 25% of leaf area affected. (After Grqaney, 1933.)

(a)   (b)   (c)   (d)

beyond the scope of this book. The aim of this chapter is to introduce the student to the key applied concepts in disease management. In reality, most growers and agronomists who are dealing with field-scale arable crops, such as wheat, will adopt a pragmatic approach to the problem and will plan a series of field walks that are carried out in a diagonal or "W" pattern across a field or a series of fields, taking samples at specific points across the fields. In practice, these sample points will be based on historic disease patterns known for the given crop and field location. Taking the leaf and stem rusts as an example, and using the key represented in Figure 10.5, the procedure for

Table 10.11 Disease assessment key for potato blight on haulm.

Symbol	Host–Pathogen Interaction
Oi	Immune: no visible signs of infection
Oc	Highly resistant: only minute chlorotic flecks
On	Highly resistant: only minute necrotic flecks
1	Resistant: small pustules with surrounding necrotic tissue
2	Moderate resistance: medium-sized pustules with surrounding necrotic tissue
3	Moderately susceptible: medium-sized pustules with surrounding chlorotic tissue
4	Susceptible: large pustules with little or no chlorosis
X	Mesothetic reaction: mixed reaction types on one leaf

(Adapted from British Mycological Society, 1947).

Table 10.12 Examples of *R* genes and the pathogens associated with their expression.

R gene	Plant	Disease	Gene Product (Protein)	Relevance of Example Plant
RPS2	Thale cress (*Arabidopsis thaliana*)	*Pseudomonas syringae*	CC-NBS-LRR	Model plant for molecular biology and plant physiology
RPS5			LZ-NBS-LRR	
RPS4			TIR-NBS-LRR	
RPM1			CC-NBS-LRR	
RPP8/HRT	Thale cress (*Arabidopsis thaliana*)	*Peronospora parasitica*	CC-NBS-LRR	
RPP13				
RPP1	Thale cress (*Arabidopsis thaliana*)	*Peronospora parasitica*	TIR-NBS-LRR	
Rpp10				
Rpp14				
RPP4				
RPP5				
SS14				
Mlo	Barley (*Hordeum vulgare*)	*Erysiphe graminis*	TM	Staple food plant and key crop for malting whisky
Mla1	Barley (*Hordeum vulgare*)	*Blumeria graminis*	CC-NBS-LRR	
Mla6				
Mla12				
Mla13				
Rpg1	Barley (*Hordeum vulgare*)	*Puccinia graminis*	Protein kinase	
L6	Flax (*Linum usitatissimum*)	*Melampsora lini*	TIR-NBS-LRR	Key fiber crop and original model for gene-for-gene theory
L, L1-L11,LH				
M				
P				
P2				
Pib	Rice (*Oryza sativa*)	*Magnaporthe grisea*	CC-NBS-LRR	Staple food plant for over 3 billion people
Pi-ta				
Pi36				
Pi9	Rice (*Oryza sativa*)		NBS-LRR	
Pi2				
Piz-t				
Xa1	Rice (*Oryza sativa*)	*Xanthomonas oryzae*	CC-NBS-LRR	
Xa21	Rice (*Oryza sativa*)	*Xanthomonas oryzae*	Kinase-LRR-ser/thr-protein kinase	
Xa21D	Rice (*Oryza sativa*)	*Xanthomonas oryzae*	LRR receptor-like protein	
Xa26			LRR receptor-like kinase	
Lr10[a]	Wheat (*Triticum* species)	*Puccinia triticina*	CC-NBS-LRR	Global staple cereal crop
Cre1	Wheat (*Triticum* species)	*Heterodera avenae*	NBS-LRR	
Cre3				

[a]*Lr10* is one example of the many *Lr* genes isolated from wheat. (Adapted from Liu et al., 2013.)

estimating disease incidence and severity and then using these estimates to support decisions about the use of fungicides can be summarized as follows:

(1) Assess the growth stage of the crop (use either Feekes or Zadoks growth stage keys).

(2) Take a random sample of fertile tillers from around 50 individual plants.

(3) Compare the samples with the disease key and assess the percentage of leaf area affected by the disease.

(4) Consider disease severity against models of economic threshold.

(a) If the disease is found on 10% of the lower leaves early in the cropping season (before heading or jointing of the stems), and weather conditions favor disease spread, a fungicide application may be advisable.

(b) If the top two leaves of plants have 5–10% infection before heading it is potentially too late to apply fungicide. If the disease is found at the late milk or dough stage and disease thresholds are low (less than 5%) and weather conditions do not favor disease spread, it may not be economically viable to apply a fungicide.

The procedure described above is for leaf- and stem-based disease. A similar approach can be applied to soil-based disease of roots and tubers, but in this case the sampling procedure is destructive. Finally, it is important to monitor throughout the life cycle of the crop to ensure the detection of early-season disease, and then to continue monitoring to maintain a record of disease progress in relation to the treatment program.

Using data derived from manual field-based assessments of disease thresholds carried out by considering a number of disease risk factors in this case study the risk factors for the brown rusts (*Puccinia triticina*) are as follows:

- High June and July temperatures together with high levels of early-morning dew

- High seasonal humidity

- Early sowing

- Emergence of new rust races that overcome the major gene resistance of current commercial varieties

- Geographical location; in the UK, moderate-risk areas range from the Midlands to Wales and Devon and Cornwall, whereas high-risk areas include the home counties, the counties of East Anglia, such as Norfolk and Suffolk, and finally Kent, Sussex, and the counties of southern Britain.

The importance of assessing and monitoring for disease cannot be overemphasized. In high-risk years (that is, years and agronomic conditions that favor the spread of cereal diseases such as the rusts and *Septoria* pathogens), it may become essential to walk the fields every 10–14 days, depending on the latency period of the pathogen. The impact on yield of foliar pathogens such as the rusts and *Septoria* species is significant. Data from CropMonitor™ (an interactive web-based platform for arable farmers in the UK, where nationwide monitoring sites across the UK report on the incidence and severity of crop pests and disease; available at www.cropmonitor.co.uk) showed that for *Septoria tritici*, leaves in 2014 had the highest levels of disease lesions since 1970, with the disease affecting 99% of the UK wheat crop. Crops had an average of 5.9% of their flag leaf covered in *S. tritici* lesions, and 11.5% of the area of leaf 2 was also covered in *S. tritici* lesions, which resulted in significant reductions in crop yield. Similar patterns were observed for *P. triticina*, with the pathogen being recorded in 7% of the UK wheat crop in 2014. Converting the above infection percentages to yield loss is relatively straightforward, but it requires the researcher or agronomist to evaluate the impact of crop resistance rating and fungicide application. Research conducted on durum wheat in Tunisia (Berraies et al., 2014) demonstrated that, on average, varieties of durum wheat susceptible to *Septoria* leaf blotch (SLB) were assessed using a categorical assessment key (rating scale of 1–9) designed by Prescott (1975). This key allows the progression of the disease to be followed from the lower leaves through to the flag leaf. In their results, Berraies et al. (2014) demonstrated average yield losses of 384 and 325 kg ha^{-1} during 2007 and 2008, respectively, for each unit of the Prescott scale. When scaled up to final untreated control, SLB disease has the potential to cause a 50% decrease in yield once it reaches the flag leaf.

Other sampling methodologies exist, and the data obtained can be used in disease forecast models. These include sampling methods such as volumetric spore traps, which intercept windblown spores such as conidia, but interpretation of the data is complex and requires simultaneous monitoring for disease expression within the crop. However, the use of volumetric spore traps provides information about the timing of spore dispersal across a region, and gives an indication of spore abundance. A new and potentially game-changing development in monitoring crops for disease and pests is the use of aerial drones. Such vehicles are being rapidly developed, and they have the ability to quickly detect differences in canopy development across a farm, which can then be employed as part of a package of precision farming techniques (see Chapter 12).

10.5 Chemical Control of Plant Disease

The control of plant pathogens by using fungicides is a historical practice that dates back to 1807, when B. Prévost introduced the use of a copper compound as a seed treatment for wheat. Copper compounds are still frequently used in modern times, and copper is one of the main constituents of the famous Bordeaux mixture that was developed by a French botany professor, Pierre-Marie-Alexis Millardet, in 1885 as a fungicide treatment for downy mildew. However, in 2014, Bordeaux mixture, along with many other pesticide compounds, was withdrawn from sale across the EU as part of an EU-wide review of pesticide safety. The arguments presented in support of the withdrawal of these pesticides were based on health and environmental safety, such as the widespread debate across the EU about the effect of neonicotinoids on the health of pollinators such as bumble bees (*Bombus* species). Neonicotinoids are a class of insecticides that are used to control flea beetle (*Psylliodes chrysocephala*) in oilseed rape crops. In 2015, the UK government relaxed the neonicotinoid ban for 120 days in areas of the UK where the flea beetle causes the highest levels of crop damage. The above-mentioned debate about pesticide withdrawals is an ongoing issue for the industry, and a cause for concern about food security issues based on our ability to protect crops from pests and diseases. There is no simple answer to this complex and socio-economic issue, but a more detailed account of the ongoing debate will be provided in Chapter 11. Setting aside the debate on the use of pesticides, it is important for students to understand the ways in which fungicides control plant pathogens, and how the agrochemical industry categorizes the compounds that it markets to growers. Basically, fungicides are classified as either contact or systemic fungicides—indeed this classification also extends to both insecticides and herbicides. Contact fungicides are compounds that control the spread of disease by preventing the germination of spores and/or directly killing the pathogen on contact. The problem with contact fungicides is that they result in incomplete protection, as it is impossible to guarantee full cover of all standing crop foliage, and contact fungicides are more vulnerable to dilution by rain. In contrast, systemic fungicides are absorbed by the plant and then translocated around it. Therefore they provide more complete protection of all plant parts, and once systemic fungicides have been incorporated into plant tissue they become rain-fast more rapidly, and hence less vulnerable to dilution effects caused by untimely precipitation. Systemic fungicides have an additional advantage in that they can also be curative in their mode of action.

Modes of Action of Fungicides

There are several active compounds available to growers, each of which provides a different mode of action against a wide range of plant disease organisms. The chemistry is complex, and a complete discussion is beyond the scope of this chapter and indeed this textbook, but a review of a range of commonly used active ingredients used in fungicide preparations will be presented based on crop species, and in cases where multiple crop types can be protected by the same active ingredient, this will be indicated.

Major Groups of Contemporary Fungicides

Chlorothalonil (CTL) is an organic compound that is used as a broad-spectrum, non-systemic fungicide on a range of crops. One of the main uses of CTL in the UK is on wheat as a protectant spray at T0, T1, and T2 (wheat spray timings are discussed in detail below), in particular to delay the development of *Septoria tritici*. Chlorothalonil is marketed under the name Bravo, and also under the names Agronil, Aminil, Bronco, and Daconil. The structure of CTL is based on a ring with two triple-bonded nitrogen atoms and four single-bonded chloride atoms. The mode of action of CTL is the reduction of intracellular glutathione molecules to forms that cannot participate in essential enzymatic reactions in the cell, ultimately leading to cellular death. CTL is available as a dust, dry and water-soluble granules, a liquid spray, a fog or a dip, and as a **wettable powder**. When applied to a crop, CTL lands on the canopy and forms a suspension of droplets in the films of water that naturally occur on leaf surfaces as a result of rain and/or dew events. The suspended CTL droplets effectively redistribute the active ingredient around the crop **phylloplane**. Thus when spores of target pathogens alight on leaf surfaces they absorb the active ingredient into their cells as part of the process of spore germination.

CTL is highly toxic to aquatic systems, and has been implicated in the decline in the Asiatic honey bee (*Apis cerana*) and western honey bee (*Apis mellifera*) populations by rendering the honey bee more susceptible to the gut parasite *Nosema ceranae*. *N. ceranae* is a microsporidian (a spore-forming unicellular parasite).

In the UK, the main use of CTL under the trade name of Bravo is as a foliar spray that provides protectant activity against a range of cereal diseases, including *Septoria* leaf blotch (*Mycosphaerella graminicola*) on wheat, rye, and triticale, as well as glume blotch (*Phaeosphaeria nodorum*) on wheat and rye, and leaf scald (*Rhynchosporium secalis*) on barley, rye, and triticale. Globally, CTL is used to control numerous foliar diseases in a range of crops, including bananas, dry beans, coffee, peanuts, potatoes, pulses, stone fruit, vegetables, tomatoes, and turf crops. Chlorothalonil is effective against pathogens belonging to the Ascomycetes, the Basidiomycetes, the imperfect fungi, and the oomycetes. It is known to be a multi-site fungicide, and consequently there have been no reported cases of resistance build-up in the target pathogens. Chlorothalonil is used as part of a tank mix, and thus can be combined with succinate dehydrogenase inhibitor (SDHI) fungicides and triazoles in the spray tank.

SDHI fungicides were discovered over 40 years ago, and now include first-generation SDHI fungicides (for example, the carboxymides) and, since 2003, second-generation SDHI fungicides (for example, bixafen, fluxapyroxad, and isopyrazam). All of these SDHI fungicides are inhibitors of succinate dehydrogenase in the mitochondrial respiration chain of numerous fungal pathogens, and hence inhibit several stages in the life cycle of pathogens, including spore germination, mycelium and germ-tube growth, and the formation of appressoria. Succinate dehydrogenase consists of four subunits (A, B, C, and D) and a binding site of the SDHIs is formed from subunits

B, C, and D. Fungal pathogens are able to overcome SDHI toxicity by mutations in the SDHI target sites, and thus resistance to widespread application of SDHI is emerging. Some examples of pathogens that are developing resistance to SDHI fungicides include *Alternaria* species (for example, *A. alternata* and *A. solani*, both of which exhibit resistance to SDHIs in the field). Other commercially important pathogens that are developing resistance to SDHI include *Botrytis cinerea* across a range of crops, *B. elliptica* on lilies, and *Sclerotinia sclerotiorum* on oilseed rape. In 2015, strains of *Septoria tritici* were reported to be developing resistance to SDHIs. This list of SDHI-resistant pathogens is not exhaustive, and careful use of SDHI fungicides is essential in order to reduce further development of pathogen resistance to this very important group of fungicides.

SDHIs are commonly used in field-scale agriculture, and are an important chemical ingredient of modern fungicide formulations such as Aviator [235]Xpro (marketed by Bayer Crop Science), in which an SDHI (bixafen) is formulated with a triazole (prothioconazole). Aviator [235]Xpro is used in wheat crops to control key diseases such as *Septoria tritici* and *S. nodorum* (leaf blotch and glume blotch, respectively) along with other key foliar and stem diseases, such as eyespot, powdery mildew, yellow rust, brown rust, tan spot, and complex ear disease, including *Fusarium* ear blight. SDHIs are also used in fungicide programs for other crops, including the control of mildew and *Botrytis* diseases in grapes and apples. SDHIs (marketed as Fontelis®, a group 7 SDHI with a single active site, by DuPont) can be used in strawberry crops to control mildew and gray mold. Fluxapyroxad, another SDHI, is a broad-spectrum pyrazole-carboxamide fungicide which is widely used on cereals, row and vegetable crops (such as peas and beans), oilseed crops (including flax, oilseed rape, and sunflower), and tuber crops (such as the potato). Fluxapyroxad is also used on a range of tree fruits, such as apples, apricots, cherries, and pears. The target organisms are the usual suspects, namely gray mold (*Botrytis cinerea*), early blight (*Alternaria solani*), and the powdery mildews belonging to the Erysiphales.

Triazoles are the largest class of fungicides, and were first launched in 1973 by Bayer Crop Science. Numerous other triazoles have been launched subsequently, the most recent compound being prothioconazole (proline), which was launched in 2004. This class of fungicides is known as demethylation inhibitors (DMIs), and they are xylem mobile and hence systemic. They have a broad spectrum of DMI activity, and consequently pathogen resistance to this group has been slow to build up, but it is now emerging among the older triazoles as a slow shift in DMI sensitivity in target pathogens. However, the newer triazoles, such as proline, are more active against their target pathogens and this has resulted in enhanced efficacy of the chemical armory.

Proline [275] (marketed by Bayer Crop Science) is widely used in arable agriculture for both cereal and oilseed rape crops. It is a broad-spectrum fungicide that is used to control a range of wheat diseases, including *Septoria*, powdery mildew, eyespot, brown and yellow rusts, tan spot, *Fusarium* ear blight (including a reduction in production of the mycotoxin known as deoxynivalenol), and the reduction of sooty molds. In addition it is used to control eyespot in rye and oat, crown rust and mildew in oats and brown rusts, powdery mildews, and *Rhynchosporium* in rye. Proline [275] is also used to control light leaf spot, phoma leaf spot, and stem canker in oilseed rape.

There are significant and growing concerns about the effects of triazole fungicides and their breakdown products on both human health and the wider environment. Triazole breakdown products such as 1,2,4-triazole, triazole alanine, and triazole acetic acid are common metabolites of the triazole groups and have been isolated in rat and mouse tissue, leading to questions about their safety in long-term use. Flutriafol ([RS]-2,

4′-difluoro-α-[1H-1,2,4-triazole-1-ylmethyl] benzhydryl alcohol) is another agriculturally important triazole with broad-spectrum efficacy in a wide range of crops. However, it is extremely persistent in the soil, where it can build up from repeated applications over time, and its residues present a high risk of mobility in soils. It is quite likely that flutriafol will find its way into ground water (Environmental Protection Agency, 1991) and, although the final fate of this compound has not fully elucidated, there are growing concerns about the impact that flutriafol will have on human health and biological diversity.

Emerging concerns about the neurotoxic effects of triazoles and their break-down products are being reported in the literature, where it is suggested that the triazole fungicide known as triadimefon, and its metabolite triadimenol, may interfere with the metabolism of the neurotransmitter dopamine.

In this chapter we have discussed just a small number of examples from the three key fungicide groups that are widely used in agricultural systems. There are many other fungicides on the market. Another important group consists of the strobilurin fungicides, which were released in 1996, and have been widely used to control numerous cereal diseases. Their mode of action involves binding to the Qo site of the cytochrome bc1 complex in fungal pathogens, and they are known as Qo site inhibitors (QoIs). However, resistance to strobilurin compounds has rapidly developed in wheat powdery mildew (*Blumeria graminis* f. sp. *tritici*), and this resistance correlates with a specific point mutation in the mildews that leads to the substitution of glycine by alanine at position 143 of cytochrome *b* (Sierotzki et al., 2000). Further and more detailed coverage of all pesticides is available in the most recent edition of *The Pesticide Manual*, published by the British Crop Protection Council.

10.6 Managing Fungal Pathogens in Wheat: A UK Model of Applied Practice

There are numerous crop production systems around the globe, and each has developed its own methodology for best practice with regard to disease management. It is beyond the scope of this chapter to review all of these systems, and instead our focus will be on the control of foliar pathogens in wheat based on the AHDB *Wheat Disease Management Guide* (https://cereals.ahdb.org.uk/media/176167/g63-wheat-disease-management-guide-february-2016.pdf). The rationale for this approach is to provide the student with a consolidated model of disease management based on a globally important staple food crop. The principles presented in the model of disease management for wheat will be applicable, with local modifications, to numerous other crop species and their suite of pathogens.

Foliar infections in wheat crops, and indeed in all cereal crops, are normally a result of spores moving into the crop, and the timing of this is disease dependent (for example, *Septoria tritici* is present on the lower leaves of wheat by early spring). Following the movement of spores into the crop, there is a latent period before symptoms develop; this is when the disease organism has germinated and penetrated the leaf, resulting in subsequent growth of mycelium. Fungicides are most effective during the latent period, including during early growth of mycelium, but once the fungus starts to develop sporulation structures, the efficacy of fungicides drops off sharply. The length of the latent period is highly variable, and is both pathogen and climate dependent. For example, *Septoria tritici* normally has a 28-day latent period, but this can shorten to 14 days in the summer, and the latent periods for brown rust and mildews can be as short as 4–5 days. Thus growers must be vigilant for early signs of disease. Foliar pathogens can also travel up the stem and infect stems and leaves higher up in the wheat canopy. As new leaves emerge they are usually free of foliar pathogens such as *S. tritici*, but as the final three leaves emerge they rapidly become susceptible to infection from vertical spread of pathogens. Hence a crop protection strategy for wheat has been developed with the objectives of delaying the onset of disease and protecting the key yield-forming leaves. This is known as the constitutional phase of crop development, which includes stem extension and the emergence of the final three leaves.

Foliar applications of fungicides to wheat have been developed to protect the key yield-forming leaves, namely leaves 3, 2, and 1 (leaf 1 is the flag leaf), and the timing of the emergence of these leaves is related to the growth stage (GS) of the crop, which has been described by a decimal coding system. Leaf 3 typically emerges at GS32, leaf 2 at GS33, and leaf 1 at GS39. The timing of fungicide sprays in relation to these key growth stages is recognized as T1 for leaf 3 and T2 for leaf 1. Optimal and timely applications of fungicides will give effective disease control for the top three leaves. However, fungicides may also be applied at T0, which corresponds to earlier growth stages, such as those that describe tillering, and a final fungicide application is referred to as T3, which corresponds to the emergence of the ear at GS59. The actual GS for leaf 3 is slightly variable, and may be as early as GS31 or as late as GS33. The optimal timing for spraying is when the leaf has just fully emerged. Spraying too early will result in poor coverage of the key foliage and thus poor suppression of foliar pathogens, and spraying too late will leave foliage vulnerable to infection from rain-splashed spores, which quickly establish on unprotected crops.

Fungicide is applied to wheat either via a tractor-mounted sprayer or using a dedicated crop sprayer. These machines have a tank-based reservoir which is pressurized so that the spray leaving the spray nozzles on the spray boom is atomized. This results in the production of a spray of very fine droplets which maximizes canopy cover of the applied product. Considerable technology underpins the use of spray nozzles and the resultant spray pattern and droplet size, but basically there are three types of sprayer designs:

- Air induction with the use of flat fan nozzles: this provides fine droplets with reduced spray drift and good to acceptable spray efficacy for application of fungicides at GS23–50.

- Conventional pressure models with two nozzle designs: the flat fan nozzle provides adequate spray efficacy for fine droplets and good spray efficacy for medium-sized droplets from the flat fan nozzle; the hollow-cone nozzle provides adequate coverage of medium-sized droplets for application of fungicides at GS23–50.

- Low-drift models: the flat fan nozzle provides medium-sized droplets; the deflector nozzle is not recommended for the use of fungicides at GS23–50.

The development of modern fungicides has enabled farmers and growers to use a complex mix of chemical compounds in the sprayer, known as the tank mix. With regard to controlling foliar and stem pathogens in wheat, tank mixes are designed to enhance the efficacy of agrochemicals and provide robust protection of key growth stages of wheat during the cropping year (**Table 10.13**).

10.7 Summary

The control of crop pathogens is a complex and evolving science that not only relies on the development of new chemical compounds to control key pathogens, but also calls upon the knowledge of the grower when deciding

Table 10.13 Example of a fungicide strategy for winter wheat grown in the south-west of the UK.

Timing	T0	T1	T2	T3
Growth stage	GS30	GS31	GS39	GS59
Leaf emergence	4	2 and 3	1 = flag leaf	Ear
Approximate date in south-west England	30 March	20 April	10 May	5 June
Strategy	Reduce *Septoria* inoculum, and control rust and mildew	Control spread of *Septoria* to leaves 2 and 3, and reduce the risk of eyespot	Crucial control of *Septoria* on leaves 1 and 2	Reduction of *Fusarium* blights
Products	Bravo and triazole fungicides	SDHI and triazole fungicides and Bravo	SDHI and triazole fungicides and Bravo	*Fusarium*-active triazole

on the best regime for treating a crop. Crop protection starts at the farm gate with rigorous biosecurity protocols and selection of modern disease-resistant cultivars, and is maintained throughout the cropping year by continuous assessment of disease pressure and timely applications of fungicides. The development of fungicide resistance and growing concerns about the negative impacts of agrochemicals are leading to a reconsideration of crop protection strategies which will in the future call upon the implementation of **integrated pest management (IPM)** procedures (see Chapter 11). Indeed, modern cereal growers already use a suite of IPM strategies to control pathogens in crops such as wheat, including variety selection, disease risk analysis, and the development of new technology and cropping systems that will enhance crop performance over the annual cropping cycle.

Further Reading

Acquaah G (2012) Principles of Plant Genetics and Breeding, 2nd ed. Wiley-Blackwell.

Agriculture and Horticulture Development Board (2017) AHDB Recommended Lists for Cereals and Oilseeds 2017/18. https://cereals.ahdb.org.uk/media/800462/ahdb-recommended-list-web.pdf

Agriculture and Horticulture Development Board (2017) Fungicide Activity and Performance in Wheat. https://cereals.ahdb.org.uk/media/1190684/lr-wheat-fungicide-is56-2017.pdf

Agrios GN (2005) Plant Pathology, 5th ed. Elsevier Academic Press.

Ambrus A (undated) Prothioconazole (232). FAO fact sheet. www.fao.org/fileadmin/templates/agphome/documents/Pests_Pesticides/JMPR/Evaluation08/Prothioconazole.pdf

Anon (2006) Potato wart disease, Fera. http://webarchive.nationalarchives.gov.uk/20090318201257/http://www.defra.gov.uk/planth/pestnote/2006/pwd.pdf

Bayer Crop Science UK. Redigo Deter. www.bayercropscience.co.uk/our-products/seed-treatment/redigo-deter/

Berraies S, Ammar K, Gharbi MS et al. (2014) Quantitative inheritance of resistance to Septoria tritici blotch in durum wheat in Tunisia. *Chilean JAR* 74(1): 35–40

Boyes DC, Nam J & Dangl J (1998) The *Arabidopsis thaliana* RPM1 disease resistance gene product is a peripheral plasma membrane protein that is degraded coincident with the hypersensitive response. *Proc Natl Acad Sci USA* 95:15849–15854.

British Mycological Society (1947) The measurement of potato blight. *Trans Br Mycol Soc* 31:140–141.

British Seed Potato Classification Scheme (undated) Agricultural and Horticultural Development Board. http://www.potato.org.uk/sites/default/files/publication_upload/british_seed_classification_schemes_%2B_graphic.pdf

Burnett F & Oxley S (2010) Potato Storage Diseases. Technical Note TN630. Scotland's Rural College (SRUC). www.sruc.ac.uk/downloads/file/747/tn630_potato_storage_diseases

Clark WS, Cockerell V & Thomas JE (2004) Wheat Seed Health and Seed-Borne Diseases: A Guide. Home-Grown Cereals Authority (HGCA). www.sasa.gov.uk/content/wheat-seed-health-seed-borne-diseases-%E2%80%93-guide

Dinesh-Kumar SP, Wai-Hong Tham & Baker BJ (2000) Structure–function analysis of the tobacco mosaic virus resistance gene N. *Proc Natl Acad Sci USA* 97:14789–14794.

Erickson FL, Dinesh-Kumar SP, Holzberg S et al. (1999) Interactions between tobacco mosaic virus and the tobacco N gene. *Philos Trans R Soc Lond B Biol Sci* 354:653–658.

EPA – U. S. Environmental Protection Agency Pesticide Monitoring, Phase II Fact Sheet Series 7 (of 14), Washington, 1991.

Faro LR (2010) Neurotoxic Effects of Triazole Fungicides on Nigrostriatal Dopaminergic Neurotransmission. http://cdn.intechweb.org/pdfs/12395.pdf

Flor HH (1956) The complementary genic systems in flax and flax rust. *Adv Genet* 8:29–54.

Granny FJ (1933) Method of estimating losses of cereal rusts. In: Proceedings of the World Grain Exhibition and Conference, Canada Jul–Aug 1933, pp. 224–236 Pub. By HW Arnold.

Hammond-Kosack KE & Jones JDG (1997) Plant disease resistance genes. *Annu Rev Plant Physiol Plant Mol Biol* 48:575–607.

Kolmer JA (1996) Genetics of resistance to wheat leaf rusts. *Annu Rev Phytopathol* 34:435–455.

Liu Y, Liu B, Zhu X et al. (2013) Fine-mapping and molecular marker development for *Pi56(t)*, a NBS-LRR gene conferring broad-spectrum resistance to *Magnaporthe oryzae* in rice. *Theor Appl Genet* 126:985–998

Liu J, Liu X, Dai L & Wang G (2007) Recent progress in elucidating the structure, function and evolution of disease resistance genes in plants. *J Genet Genomics* 34:765–776.

McDowell JM & Woffenden BJ (2003) Plant disease resistance genes: recent insights and potential applications. *Trends Biotechnol* 21:178–183.

Saari EB & Prescott JM (1975) A scale for appraising the foliar intensity of wheat diseases. *Plant Dis. Rep* 59:377–380

Sierotzki H, Wullschleger J & Gisi U (2000) Point mutation in cytochrome *b* gene conferring resistance to strobilurin fungicides in *Erysiphe graminis* f. sp. *tritici* field isolates. *Pesticide Biochem Physiol* 68:107–112.

Slawson D (2008) Turning Over a Clean Leaf: How to Protect Your Garden from Pest and Disease Invaders. http://plantnetwork.org/wordpress/wp-content/uploads/0/parksgardensbiosecurityposter.pdf

Slawson D (2013) Turning Over a Clean Leaf: How to Protect Trees from Pests and Diseases When Working in Woodlands and Forests. www.forestry.gov.uk/pdf/Poster_Forestry_Biosecurity_2013. pdf/$FILE/Poster_Forestry_Biosecurity_2013.pdf

US Environmental Protection Agency (1991) Pesticide Monitoring. Phase II Fact Sheet. US Environmental Protection Agency, Washington, DC.

Van Gastel AJG, Bishaw Z & Gregg BR (2002) Wheat seed production. In Bread Wheat: Improvement and Production (Curtis BC, Rajaram S & Gomez Macpherson H eds), pp. 463–464. Food and Agriculture Organization of the United Nations.

Yang S, Li J, Zhang X et al. (2013) Rapidly evolving *R* genes in diverse grass species confer resistance to rice blast disease. *Proc Natl Acad Sci USA* 110:18572–18577.

Useful Websites

Department for Environment, Food & Rural Affairs, and Animal and Plant Health Agency (2017) Guidance: Fruit Propagation Certification Scheme. https://www.gov.uk/guidance/fruit-propagation-certification-scheme#history

Chapter 11
Organic and Integrated Methods

Chapter 10 established the key principles underlying the management of plant pathogens, with an emphasis on biosecurity and chemical control options. However, over the last 40–50 years there has been growing concern expressed by numerous interested parties about the effect of agrochemicals on both human health and the environment. These concerns, coupled with continued development of fungicide-resistant pathogens, have led to a shift in disease management philosophy away from over-reliance on fungicides and towards the development of new cropping systems that rely on either integrated pest management (IPM) techniques or exclusively organic systems. These approaches endeavor to either reduce or completely remove the need to use synthetically produced pesticides. The crop production strategies in both integrated and organic systems are based on the principle of a mixed, healthy cropping environment in which complex cropping rotations and mixtures of crop species encourage a diverse array of organisms within that environment, some of which will be natural predators of the pest species, while others will of course be the antagonists (the pathogens in this case). The pathogens are controlled by avoiding the monocultures that are commonly observed in large-scale arable systems. Monocultures favor the development of widespread epidemics, based on the build-up of large populations of inoculum over a large cropping area, whereas integrated pest management strategies and organic systems rely on diversity and rotations which break the cycle of species-specific pathogens and thus aim to reduce sources of inoculum, both spatially and temporally. Modern IPM and organic systems are complex and often very technical. In particular, soft fruit and salad crops rely on modern, covered cropping systems, where the cropping environment is centrally managed by a computer.

Organic systems start their production cycle with attention to soil fertility, and include soil fertility-building crops such as alfalfa (*Medicago sativa*), a legume that belongs to the family Fabaceae. These soil fertility-building crops will have their own suite of pathogens, and then, following the soil fertility-building step, organic growers implement a series of crop rotations to avoid the development of diseases in the second crop. In most cases, organic systems are based on the mixed farming model and therefore will include livestock on the farm, such as cattle, sheep, pigs, and chickens. The waste from the livestock system is then used as part of the soil fertility-building phase in organic systems, and if the livestock system includes beef and sheep, a grass ley will form part of the crop rotations. Some IPM systems are solely arable, and in these systems the IPM element is a mixture of crop rotations, cultivar mixes, and complex tank mixes of fungicides that have multi-site activity.

(a)

(b)

(c)

(d)

Figure 11.1 A range of soil fertility-building crops. (a) Grass–clover ley; (b) vetch; (c) sainfoin crop; (d) flower of sainfoin. A color version of Figure 11.1b can be found in the color plate section at the end of the book.

This chapter will provide an overview of issues associated with organic production systems, and will also explore the concept of integrated disease management in both covered crops and conventional arable cropping systems.

11.1 Organic Production Systems

Soil Fertility Building

Organic farming is a production system based on sustaining healthy soils and regional ecosystems. Organic production is based on the central principle of food production without the input of synthetically derived agrochemicals and inorganic artificial fertilizers. Organic farming relies on natural systems, rather than trying to dominate them (as in conventional cereal production), and to achieve this organic farmers endeavor to encourage and enhance the biological cycles that are inherent across the farm and in particular within soils. These farmers view the production environment holistically, and try to strike a balance between the production environment and the natural landscape. They start their production system by building soil fertility, which begins with the establishment of grass leys and soil fertility-building crops. These crops are typically based on the legumes (the family to which these species belong is now known as the Fabaceae, but most farmers still refer to these crops as legumes, and therefore this volume will do likewise), such as alfalfa, clovers, sainfoin, and vetch (**Figure 11.1**). Soil fertility building is a slow process, and will also benefit from inputs of composted organic manure. The aim of the soil fertility-building phase is to improve the structural and organic matter content of the soil, and thus soil fertility, particularly with regard to key crop nutrients such as nitrogen, phosphorus, and potassium. If a farm that is converting to organic production was traditionally based on a mixed farming model (that is, arable and livestock), the conversion process is not so problematic, as the farm would have used organic waste from its livestock as part of the annual crop nutrient budget, and of course it will have a ready supply of farmyard manure to facilitate conversion to organic systems.

However, if the farm that is converting to organic production was historically a modern conventional cereal farm (that is, with no livestock), the soil fertility-building phases are crucial. This is because long-term conventional cereal production has, over time, resulted in large losses of soil carbon, and hence soils on many conventional cereal farms are structurally poor and have very low nutrient-holding capacity. Many soils on conventional cereal farms will also have large overwintering populations of plant pathogens residing on crop debris. This is because many modern cereal farms use minimum-tillage systems (**Figure 11.2**) to cultivate their soil, as these systems incorporate the crop debris into the surface soil horizons, which helps to increase organic carbon content and stabilize poorly structured soils. However, this also leads to high carry-over levels of pathogen inoculum, such as resting spores, on crop debris. The conversion from conventional cereal production to organic production is a lengthy process, and requires a shift in thinking and the adoption of numerous traditional approaches, including new crop rotations and rotating soil cultivation techniques. The traditional plough will have a role to play in organic production systems in that ploughing can help to manage the **green bridge** (**Figure 11.3**). In some crop rotations the green bridge will need to be buried in order to reduce the incidence of pathogens on crop debris. Deep ploughing (**Figure 11.4**) will both bury crop debris and help to reduce weed populations in surface soil horizons between rotations, which can be quite problematic in organic systems (**Figure 11.5**).

Managing Disease in Organic Crops

Protecting crops from plant pathogens in an organic system is not straightforward, but it is possible to produce a healthy and relatively disease-free

(a)

(b)

Figure 11.2 Minimum-tillage equipment. (a) A set of discs and tine-mounted discs. (b) The effect of minimum-tillage cultivation on the surface soil horizon, illustrating how the stubble is incorporated into the surface soil horizon, which stabilizes surface soils but means that crop debris is left within the crop emergent layer, which is thus vulnerable to any carry-over inoculum.

crop by rotating crops around the farm in such a way that populations of plant pathogens do not become major problems. The best approach to this is of course to adopt a mixed farming model, but for some organic growers this is not a viable option, and therefore more attention has to be given to soil fertility-building rotations than would be the case in a mixed farming system. The key principle in crop rotations is to mix up the types of crops so that no one plant species dominates the farm. This can be problematic for cereal growers who rely on wheat as the main component of the cropping year. There are ways to overcome this difficulty—for example, rotating to either an oilseed rape crop or an oat crop following the wheat crop. In this way these farmers can mitigate against major outbreaks of wheat disease such as *Septoria tritici*. However, this is neither straightforward nor completely without risk. These farmers would need to rely on cultivar

Figure 11.3 The green bridge. This photo shows how crop regrowth occurs through the previous stubble and is thus vulnerable to carry-over inoculum from the previous crop.

diversification schemes and sow a range of disease-resistant cultivars across the farm to help to reduce the risk of major outbreaks of *S. tritici*. However, the pathogen can often overcome the resistance mechanisms that have been bred into new cultivars. Basically, crop rotations are the next critical phase in managing disease in organic systems, following on from soil fertility building. Furthermore, well-designed rotations complement the development and maintenance of soil fertility.

Crop Rotations and Intercropping

The development of crop rotations in Europe dates back to at least the Roman times, and in the UK the well-known three-course rotation, namely autumn-sown wheat → spring wheat → fallow, dates back at least 1500

Figure 11.4 Traditional ploughing techniques. This photo shows how the plough buries the crop debris and thus reduces carry-over inoculum.

Figure 11.5 Weeds within a red cabbage crop. The weeds can act as secondary hosts of many pathogens. A color version of this figure can be found in the color plate section at the end of the book.

years. However, in the eighteenth century the use of root crops (of which turnip was the most popular at that time) and clover began to hold sway on British farms. This facilitated the development of a four-course rotation, namely clover → wheat → barley → root crops. It was then possible to omit the fallow. This new four-course rotation became the mainstay of agriculture for many decades. The four-course rotation could be manipulated and the clover replaced with a grass ley and/or other legumes. The rapid expansion of the human population brought an end to the four-course rotation at the beginning of the twentieth century, and indeed ushered in the adoption of modern commercial arable production systems, which eventually led to the modern industrialized monocultures that rely on agrochemical inputs to suppress and control plant pathogens.

The four-course rotation lends itself to modification and is ideally suited to a progressive mixed organic farm model, as the rotation can be developed to include a range of crops that have commercial markets and/or allow the growing of wheat cash crops. Such farmers could modify the traditional four-course rotation into a six- or even nine-course rotation as follows:

root crops → barley → clover/ryegrass ley → potatoes → wheat → oats.

The principle of this rotation is a mixture of broadleaf crops, followed by a monocotyledon crop (barley), then a switch to a soil fertility-building crop, followed by a broadleaf crop, and then 2 years of cereals with wheat following potatoes, as the use of potatoes results in the burial of the green bridge and a fine seed bed, which are key parameters for high-yielding wheat crops. Oats follow wheat, as oats are not susceptible to the major diseases of wheat.

In modern agriculture one of the basic principles of organic farming is to break up the cereal monocultures by introducing broadleaf crops and forage crops. This enables growers to manage weed and disease pressure and simultaneously build soil fertility. A final comment on rotations concerns the introduction of oilseed rape into modern agriculture, and the ability of growers to incorporate this broadleaf, combinable arable crop into their business model. This can be further complemented if livestock are part of the farm business model, in which case a soil fertility-building and grazing crop can be introduced as follows:

grass ley → legume → cereal (such as wheat) → oilseed rape/kale → cereal whole crop → grass ley

or

legume → oats → cereal whole crop → oilseed rape/kale → grass ley/legume.

The above two examples allow growers to move disease-susceptible crops such as wheat and oilseed rape around the farm and thus reduce continuous cycles of diseased crop residues, which are buried once the grass ley has been ploughed out.

Some key principles of crop rotations are summarized below:

- Nitrogen-fixing crops should be alternated with nitrogen-demanding crops.

- Catch and green manure crops and undersowing of crops should be used to maintain vegetative cover during winter. This reduces weed populations, nutrient leaching, and soil erosion.

- Shallow-rooting crops should be followed by deep-rooting crops. This helps to maintain soil structure and to keep soils free draining.

- Use crops in the rotation that have high root biomass, as this provides niches for soil organisms such as earthworms, thus maintaining soil health.

- Switch between straw and leaf crops, as this aids weed control.

- Incorporate spring-sown crops, as this redistributes workloads and can help to maintain disease-free periods, as the spring-sown crop will not have been subjected to the vast array of winter pathogens.

- Adopt a cultivar diversification program across the key cash crops, such as wheat and oilseed rape.

- If wheat is still a major component of the farm business model, some consideration should be given to older long straw varieties, as these tend to root deeper and have a more closed canopy, which helps to suppress weeds.

Many organic producers focus on vegetable crops, in which case there is no cereal in the rotation. This production system has its own specific problems, as common diseases such as the rusts can devastate vegetable crops, such as the leeks, and disease pressure can become crop wide (**Figure 11.6**). In this case, rotations become an important component of the system, and if this can be coupled with livestock, the incorporation of forage crops and grass leys helps to break up rotations and reduce the effect of diseased crop debris on the following crop. One of the key problems with regard to organic vegetable growing is the development of dense weed populations among the crops, which need to be controlled, as many weed species can be secondary hosts for a range of diseases, such as light leaf spot of brassicas. Management of these weeds is challenging in organic systems, and relies on hand weeding, rotations, and even burning off of germinating weeds.

In smaller production systems, intercropping (**Figure 11.7**) increases diversity across the production area, and if the intercropping is carefully planned the sequence of crop harvest dates can be planned and crops can be sown so as to avoid the major disease window (for example, early-grown crops such as *Vicia faba* can be brought to harvest before key pathogens such as the rusts

(a)

(b)

Figure 11.6 Leek rusts. (a) Field leeks showing widespread distribution of leek rusts within the lower leaves of the crop; the yellow leaves are covered in rust pustules. (b) Close-up view of rust pustules on leek leaf. A color version of Figure 11.6b can be found in the color plate section at the end of the book.

Figure 11.7 Intercropping with a mixture of crop types and the use of drip-line irrigation (the black pipes running across the soil), which can be used to fertigate crops with compounds such as liquid seaweed extract.

have the opportunity to compromise the harvest). Early crops can be established as young plants in covered structures (Figure 11.8) and sown out as soon as seasonal conditions are optimal for crop growth, thereby facilitating crop establishment and the development of healthy plants, which improves their ability to withstand disease. If these crops are pre-planned, water fertigation systems can be incorporated into the field (Figure 11.7), and organic crop stimulants such as liquid seaweed extract (LSE) can be applied to the root zone of the crops to enhance root establishment, crop growth, and the innate ability of the crop to withstand plant pathogens. LSE is widely used by organic growers, and indeed by many conventional growers, as regular low-dose applications of LSE can enhance root growth and innate physiological defense mechanisms in many crop species.

Intercropping is the principle of mixing crop species within a plot. This can be as simple as growing allium species next to other crops, such as leafy vegetables, as the exudates from allium species have antibiotic potential.

Figure 11.8 Young plants being raised in a covered structure for transplanting into crop beds later in the cropping season.

Other examples of the use of companion crops include growing legumes in and among corn crops such as sweet corn; this is a very old practice based on traditional food production by Native Americans in the USA. The basic principle of intercropping is very similar to that of crop rotations, namely to reduce the land area covered by a single crop species. However, in the case of intercropping, the mixture of crops is within one cropping field.

Organic Fungicides and Plant Biostimulants

Organic growers are not without some form of armory to fight plant pathogens. They are allowed to use elements such as sulfur (which is known to help to mitigate diseases such as the mildews and botrytis), as well as copper, oils, and bicarbonate. However, some caution should be exercised, as pesticide regulations may apply if compounds are mixed or artificially derived, and copper-based compounds have had their license withdrawn in Europe. The oils, such as neem oil, are used mainly to control insects, which of course are vectors for many plant viruses, such as mosaic viruses and potato Y virus. However, care is needed when using oils, as certain plants are sensitive to these compounds and can be easily damaged by them, including walnut (*Juglans nigra*), the ornamental Japanese maples (*Acer palmatum*), and many ornamental conifers, such as the junipers. Thus the fact that a fungicide compound is derived from an organic source does not necessarily mean that it is safe to use on all plants and all crops. Another problem is that these organic fungicides will have a last-use withdrawal period before harvest, just like commercial chemical fungicides.

Current research is investigating the potential for using suspensions of antagonistic fungal organisms against a range of plant pathogens. For example, the antagonistic yeast (*Candida diversa*) combined with plant elicitors such as harpin has been shown to suppress post-harvest pathogens such as *Botrytis cinerea* and *Penicillium expansum* in kiwifruit (Tang et al., 2015). Another promising biofungicide is based on *Coniothyrium minitans*, which targets *Sclerotinia* species in a wide range of dicotyledonous crops. The application of microorganisms does not have to be targeted against a pathogen to be effective in enhancing plant health. It has long been recognized that a range of *Trichoderma* species enhance nutrient uptake in plants but can also be antagonistic against plant pathogens. Such applications will require considerable development before they become commercially viable in a field situation, but they offer a new approach to disease management.

The use of an elicitor in the above example is a key concept in plant disease management, and applications of LSE have a similar effect. Trials using LSE as a foliar spray on greenhouse-grown carrots (*Daucus carota* subsp. *sativus*) have demonstrated that applications of LSE reduced plant infections with *Botrytis cinerea* for up to 25 days post LSE application, and were also more effective in reducing *B. cinerea* infections than applications of the well-known plant defense elicitor salicylic acid (SA) (Jayaraj et al., 2008). The mechanism of action of both LSE and SA was elucidated, and it was found that both of these compounds initiated innate plant defense mechanisms, such as production of defense-related enzymes and higher transcripts of pathogenesis-related proteins, chitinases, and other pathogenesis-related gene transcripts and their gene products.

11.2 Integrated Pest Management

The origins of integrated pest management (IPM) are more traditionally aligned with management of crop pests (for example, invertebrates such as aphids and weevils). However, IPM has a role to play in the management of plant pathogens, but in this case it is used to control the release of new disease-resistant cultivars and the application of key fungicides in an attempt

Figure 11.9 A typical layout for a cereal variety field trial.

to stem the development of fungicide resistance and the breakdown of disease resistance in new crop cultivars. The central principle is not too dissimilar to that of crop rotations, in that the grower will use a range of cultivars across the farm and even across a cropping region where farmers work cooperatively. This is then combined with careful use of fungicides to reduce the overuse of single-site active compounds and the adoption of tank mixes and multiple-site active compounds. A good example of management of a commercial crop using IPM strategies concerns wheat and the threat posed to this crop by *Septoria tritici*.

Cultivar development is an ongoing process involving many years of plant breeding (Chapter 10), field trials of new cultivars (**Figure 11.9**), and harvest evaluation of these field trials (**Figure 11.10**), where harvest data are compared with extant cultivars and the yield data are measured as either untreated or fungicide treated. The final outcome will be the release of a new cultivar that has increased resistance to, say, *S. tritici*, which if adopted

Figure 11.10 Harvesting a variety field trial plot. Note the block design (the small plots delineated by the lines in the photo) and the use of a small plot harvester, which separates out each variety for yield analysis.

Table 11.1 Summary of National Institute of Agricultural Botany (NIAB) wheat groups.

NIAB Group	Description
1	Varieties that produce consistent milling and baking flour
2	Varieties that have bread-making potential but not consistent grist
3	Soft endosperm varieties that are suited to biscuit and cake flour
4	Varieties that are grown as feed wheats

by all wheat growers within a region, would potentially see resistance to *S. tritici* break down within just a few years as this widespread adoption of a single new cultivar imposes a strong selection pressure on the target pathogen. Thus the IPM approach would be to adopt a range of wheat-growing strategies whereby the grower would, or could, grow a range of wheat groups (Table 11.1) and select a suite of cultivars with different disease ratings for the key wheat pathogens for use on their farm (Table 11.2). They would then combine this with careful rotations such as:

wheat → oats → oilseed rape → wheat.

This gives 2 years of wheat in 4 years of cultivation, and introduces a broad-leaf crop among the cereals. This could be further developed to introduce a legume-based crop.

Then, during the wheat-growing years, careful applications of key fungicides will stem the development of fungicide resistance and the breakdown of disease resistance in the new cultivar, and the rotation will help to reduce inoculum levels on crop debris.

11.3 Summary

The adoption of an organic production philosophy does not mean that growers cannot produce key cereals and/or other food crops, as they have a range of tools in their armory to assist them. A word of caution is needed, as some pathogens (such as the ergots) are very hazardous to human health and thus

Table 11.2 A shortened version of the Home-Grown Cereal Authorities (HGCA) recommended lists for winter wheat varieties for 2017–2018.

Variety	KWS Zyatt	Skyfall	Crusoe	Gallant	KWS Siskin	KWS Barrel	Spyder	Bennington	Savello
NIAB group	1	1	1	1	2	3	3	4	4
Recommended region	UK						East and West		North
Fungicide-treated grain yield (% treated control)									
UK (10.7 t/ha)	102	101	97	97	103	103	100	104	103
East region (10.8 t/ha)	102	100	97	97	103	102	101	105	104
West region (10.7 t/ha)	103	101	98	96	104	102	101	103	101
North region (9.9 t/ha)	[98]	101	93	93	101	109	97	[101]	[106]
Endosperm texture	Hard					Soft			Hard
Disease resistance rating (scale of 1–9, where 9 = high disease resistance)									
Mildew	7	6	7	6	9	6	9	7	7
Yellow rust	7	6	9	4	9	8	6	7	8
Brown rust	6	9	3	7	5	6	7	7	4
Septoria tritici	6.4	6.0	6.7	4.6	6.8	4.4	5.7	6.2	5.3

[ ] denotes limited data.
The table illustrates the disease resistance scoring and yield data compared with treated controls.

need to be carefully monitored in organic systems. Other pathogens (such as the *Fusarium* species) produce toxins that are again quite deleterious to humans and livestock, and which cannot easily be controlled in organic production systems. Finally, even if organic production is not an attractive option to a modern cereal grower, it is becoming increasingly common for these conventional growers to adopt IPM strategies in the fight against key crop pathogens that are continually evolving defense mechanisms in response to our armory of tools for combating them.

Further Reading

Bottrell DG & Smith RF (1982) Integrated pest management: this economical approach to environmentally sound pest control has dim prospects for increased federal support. *Environ Sci Technol* 16:282A–288A.

Burchett S (2000) The mechanism of action of liquid seaweed extract in the manipulation of frost resistance in winter barley (*Hordeum vulgare* L.). PhD thesis, Department of Agriculture and Food Studies, University of Plymouth.

Burchett S & Burchett S (2011) Mixed farming. In Introduction to Wildlife Conservation in Farming, pp. 19–76. John Wiley & Sons Ltd.

Glare T, Caradus J, Gelernter W et al. (2012) Have biopesticides come of age? *Trends Biotechnol* 30:250–258.

Jayaraj J, Wan A, Rahman M & Punja ZK (2008) Seaweed extract reduces foliar fungal diseases on carrot. *Crop Protect* 27:1360–1366.

Lampkin N (2002) Organic Farming. Old Pond Publishing.

O'Driscoll A, Kildea S, Doohan F et al. (2014) The wheat–*Septoria* conflict: a new front opening up? *Trends Plant Sci* 19:602–610.

Tang J, Liu Y, Li H et al. (2015) Combining an antagonistic yeast with harpin treatment to control postharvest decay of kiwifruit. *Biol Control* 89:61–67.

Chapter 12
The Future of Plant Pathology

Plant pathology is a complex, dynamic, evolving discipline. Here we have endeavored to introduce this topic to students in a tangible and visual manner. In Section 1, we covered the principles of plant pathology, gave an overview of plant pathogen types, introduced infection processes and host responses to infection, and considered the field-scale spread of these diseases. The compendia in Section 2 provided a representative list of plant pathogens with brief descriptions, and in Section 3 we have looked at how these diseases are managed. In this book, then, we have considered what plant pathogens do, what they are, and how we deal with them.

Now let us return to the interactions between pathogen and host. The pathogen infects the host, the host has an armory of responses, the pathogen then develops counter-responses, and the host responds to these with counter-counter-responses, and so on. This is a wholly natural set of processes—an "arms race" that has raged since time immemorial—and as plant pathologists, technologists, and growers all we can really hope to do is to give the plant host the edge in this arms race. We can tackle this in a number of ways, including developing resistant cultivars, gaining an understanding of the genetics of both parties in the host–pathogen relationship with a view to genetic manipulation, precision farming to maximize growing and input efficiency, post-harvest technology development (to ensure, in the face of growing demand, that maximum yields remain intact all the way to the end user), and responding to the often rapid changes in climate, and thus the suite of pathogens to be tackled, together with the crops that best suit the changing climatic conditions.

12.1 Apples

Consider the humble apple! If you drop it, it becomes bruised and damaged. If you eat it straight away, it will be fine, but by the next day large areas will have become inedible, and within a week it will be unusable. If you don't drop it, it will last for weeks in the fruit bowl and will still be good to eat. So let us now turn the clock back and consider how this apple found its way into your fruit bowl.

- The apple variety was bred for taste, color, texture, longevity, and disease resistance.

- It grew in an orchard, subjected to an ongoing barrage of pathogens, pests, and variable environmental conditions from the early stages of growth to fruit development, and in order to combat these challenges it was sprayed at least 15 times, usually more.

- The apple was harvested by machine along with thousands of others, and was then transported to a processing plant.

- It was washed, underwent scrutiny on a conveyor belt, was treated to prevent post-harvest infections, and was then waxed.

- It was packed in a box and placed in cold storage for up to 9 months.

- It was then re-checked for quality and damage and transported to a distribution center.

- There it was selected and transported again to a retailer, where it was put on display to customers.

- It was probably picked up and put down again by a number of potential buyers.

- Finally it was selected, put in a shopping trolley, probably had other items dumped on top of it, and was then purchased and taken to the customer's home in a bag, in which it rubbed up against various packets and cans before it finally found its way into the fruit bowl.

Thus the grower, the processor, the packer, the storer, the transporter, and the supermarket all play a role in ensuring that your apple reaches you in bright, shiny, fresh, and unblemished condition!

The technology associated with each of the processes listed above is constantly being improved, so let us consider some of these processes in a little more detail. Plant breeding selects for a range of qualities, including improved taste, good texture and color, storage longevity, and resistance to such diseases as *Venturia inaequalis*, a common fungal disease that causes scabbing on apples (Figure 12.1). This is not a simple process. Many crosses fail to yield all of the qualities that are being looked for, as selection for good quality in one trait frequently deselects for the desirable qualities of another trait, due to the fact that all such traits are genetically encoded. Long gone are the days when plant breeding was a process of "try it and see", so a comprehensive genetic knowledge base is key to successful breeding, both for selection of desirable traits and for ensuring that those traits are sufficiently genetically robust to be conserved across successive growing seasons.

Our apple is grown in the open in an orchard with many other apple trees around it—a monoculture. Monocultures are particularly susceptible to pests and pathogens as they represent a large host base. For the grower, damaged and diseased fruits have no value, and in fact they incur costs as they require labor-intensive attention, since infected fruits are a source of infection for the rest of the crop. Organic production notwithstanding, apple crops in particular require frequent spraying to protect them from pests and diseases, and timing is critical as the grower wants to eradicate the offending organism before any damage or pathogen replication can occur, while

(a)   (b)

Figure 12.1 The Braeburn apple (*Malus domestica* var. "Braeburn") is a popular variety sold in the supermarkets. (a) shows resistant cultivar and (b) shows susceptible cultivar. A color version of this figure can be found in the color plate section at the end of the book.

at the same time avoiding causing harm to pollinators. However, different pests and pathogens affect the crop at different growth stages, and under different environmental conditions, so ongoing treatment with cocktails of insecticides and fungicides is required. Again, the required technology is at hand. Improving the efficiency and timing of sprays, in addition to growing crops that have been bred for disease resistance, helps to reduce inputs. Today the consumer is far more likely to ask questions about the chemicals that have been used on their food, so consumer pressure is another factor that growers have to take into consideration.

Historically, harvesting was always a manual process, and on a small scale this is still the case. However, the apple industry is vast, and hand-picking apples on a large scale would clearly be unacceptably time-consuming and expensive in terms of labor costs. Modern apple-harvesting equipment has been designed to maximize protection of the fruits. This equipment varies in design, depending on factors such as orchard layout and row access. It includes tree shakers to release the fruits, and either front or side sweepers to gather the crop gently. Many apple-harvesting machines have inbuilt cleaner systems, and conveyors to deliver the apples to the trailer for transport back to the sheds, where they can be checked and packed.

The fruits will continue to carry out normal metabolic processes (that is, respiration and aging processes). For storage purposes, although metabolism cannot be stopped altogether, lowering metabolic rates increases the potential storage period. Cooling to temperatures just above freezing, together with a reduction in pressure (hypobaric treatment), slows down all metabolic reactions, enabling apples to be stored for long periods. During storage the fruits may become susceptible to a range of environmental and/or pathogenic post-harvest conditions, as physical damage to the fruit provides an entry route for pathogens (Table 12.1).

Of course, in modern times, apples can be purchased from all around the world. In the UK, for example, New Zealand apples can be seen on our shelves throughout the summer months until the new-season UK crops become available. This means that transportation is a key factor in the apple market, and can involve thousands of food miles.

Table 12.1 Post-harvest conditions in apples.

Condition	Symptoms	Cause
Bitter pit	Brown spotting throughout the flesh	Calcium deficiency
Brown heart	Browning of the core	Long-term storage damage
Fungal infection	Gray mold	*Botrytis cinerea*
Fungal infection	Blue mold rot	*Penicillium expansum*
Fungal infection	Lenticel rot	*Phlyctaena vagabunda*
Internal cork	Brown spotting throughout the flesh	Boron deficiency
Low temperature breakdown	Browning of the cortex tissue	Storage temperature too cool for too long, water content too high
Senescent blotch	Gray blotchy areas caused by long-term storage	Long-term storage damage
Senescent breakdown	Browning of the flesh	Long-term storage damage
Soft scald	Softening and darkening of the flesh	Long-term storage damage
Superficial scald	Browning on the surface of the fruit	Long-term storage damage

12.2 Food Miles

Today we find nothing unusual about the presence of apples (a commonly grown UK crop) from New Zealand on our supermarket shelves. However, not so many years ago we would have expected to have salad crops only in the summer, strawberries only in June, other berry fruits such as raspberries only in late summer, and orchard fruits from late summer onwards for a few months. Now we expect all of these crops and a whole range of exotic fruits and vegetables to be available all year round, provided by a combination of international growers and mind-boggling transport logistics. What is on the plate at dinnertime is often far more well travelled than its consumer, whether it is French beans from Kenya that will have travelled well in excess of 4000 miles to reach the UK, bananas from Costa Rica that have travelled over 5000 miles, or apples from New Zealand that have travelled around 13,000 miles. Of course not only the UK but also many other countries around the world both import and export foodstuffs on a large scale, and each of these crops comes with its own suite of pathogens. Table 12.2 lists the top 30 global crops that are transported around the world, each of which is accompanied by the potential threat of new suites of pathogens that may or may not survive the transition to a new environment, and may or may not adapt to newly available hosts at their destination. One example of this potential threat is the Ug99 strain of wheat stem rust (*Puccinia graminis* f. sp. *tritici*). First identified in Uganda in 1999 (hence the name), this disease has so far spread across many African and Middle Eastern countries, and is responsible for devastating wheat crop losses. In the UK, we import around 40% of our food, and with each pallet load comes the potential for major food security issues. Once again, technology and legislation must be relied upon to ameliorate these threats.

12.3 Import and Export of Plant Material

There are strict guidelines in place to ensure that foodstuffs and nursery stock which are being imported to or exported from the EU do not carry pathogens or pathogen vectors, and government bodies continually police all plant materials that enter or leave the EU. For example, in December 2015 the EU banned the import of tomatoes, peppers, and aubergines from Ghana, for review after a period of 1 year, due to the risk that they carry with them non-indigenous whitefly, fruitfly, and thrips due to poor phytosanitary procedures, an audit carried out in September 2016 resulted in an extension of this ban for a further year. The risk of importing exotic pathogens with these vectors cannot be overstated. In addition, new stringent laws have been introduced to prevent the import of nursery plant stock that could carry *Xylella fastidiosa* (see Chapter 7). This disease affects olive trees in particular, although it has a broad host range, including other trees, shrubs, and herbaceous plants. Importers are required to ensure that plant imports from areas where *X. fastidiosa* is known to be present have plant passports for bringing in any potential plant host. These are just two examples that have been highlighted recently, but many more potential threats undoubtedly exist. In the UK, to avoid confusion the Department for Environment, Food and Rural Affairs (Defra) has issued a clear set of guidelines to help growers and wholesalers to understand the regulations associated with import and export of plant material.

12.4 Global Climate Change

This is a particularly topical issue (it is in the news, much talked about in schools and universities, and there are frequent television documentaries about it), but when the topic of global climate change is raised in everyday conversation, understanding of it among many people seems to be patchy at

Table 12.2 Top 30 global crops with representative examples of associated diseases.

Crops in Alphabetical Order by Common Name	Disease Examples			
	Fungi	Bacteria	Viruses	Other
Apple (also pear, quince)	*Alternaria, Gymnosporangium, Monilinia, Venturia*	*Agrobacterium, Erwinia, Pseudomonas*	*Ilarvirus* species (apple mosaic virus), *Nepovirus* species (tomato ringspot virus), *Trichovirus* species (apple chlorotic ringspot virus)	*Phytophthora* species (various rots), *Podosphaera leucotricha* (apple powdery mildew)
Banana	*Armillaria, Chalara, Cladosporium, Fusarium, Verticillium*	*Erwinia, Pseudomonas, Xanthomonas*	Banana bract mosaic virus, banana bunchy top virus, cucumber mosaic virus	
Barley	*Alternaria, Claviceps, Erysiphe, Fusarium, Pseudoseptoria, Puccinia*	*Pseudomonas, Xanthomonas*	Barley yellow dwarf virus, barley yellow mosaic virus, *Bromovirus, Closterovirus*, northern cereal mosaic virus	*Pythium*
Brassicas	*Alternaria, Fusarium, Rhizoctonia, Sclerotinia*	*Pseudomonas, Xanthomonas*	Beet curly top virus, cauliflower mosaic virus, turnip yellow mosaic virus	*Pythium*
Carrot	*Alternaria, Botrytis, Erysiphe, Fusarium, Rhizoctonia*	*Agrobacterium, Erwinia, Xanthomonas*	Carrot mottle virus, carrot necrotic dieback virus, carrot red leaf virus, cucumber mosaic virus	Aster yellows phytoplasma, *Phytophthora, Pythium*
Cassava	*Armillaria, Fusarium, Glomerella, Sclerotium, Uromyces, Verticillium*	*Agrobacterium, Erwinia, Xanthomonas*	African cassava mosaic virus, cassava green mottle virus, cassava brown streak disease	*Phytoplasma*
Citrus (oranges, lemons, limes)	*Alternaria, Aspergillus, Fusarium, Ganoderma, Mucor, Penicillium, Rhizoctonia*	*Canditatus[a], Pseudomonas, Xanthomonas, Xylella*	Citrus psorosis virus, citrus tristeza virus, citrus variegation virus, satsuma dwarf virus	*Phytophthora, Pythium*
Cocoa	*Armillaria, Fusarium, Glomerella, Macrophoma*	*Agrobacterium, Erwinia*	Cacao necrosis virus, cacao swollen shoot virus	*Cephaleuros virescens, Phytophthora*
Coffee	*Armillaria, Cercospora, Colletotrichum, Fusarium, Glomerella*	*Pseudomonas*	Blister spot virus, coffee ringspot virus	*Phytomonas*
Cotton	*Alternaria, Fusarium, Glomerella, Macrophomina, Puccinia, Verticillium*	*Erwinia, Pseudomonas, Xanthomonas*	Cotton leaf curl virus	*Pythium*
Cucurbits	*Botrytis, Choanephora, Cladosporium, Fusarium, Sclerotinia, Septoria, Verticillium*	*Erwinia, Pseudomonas, Xanthomonas*	Cucumber mosaic virus	*Phytophthora, Pythium*
Currants	*Botrytis, Cronartium, Drepanopezizia, Puccinia, Sphaerotheca*	*Pseudomonas*	Blackcurrant reversion virus, currant mosaic virus, tomato ringspot virus	*Phytoplasma*
Date	*Alternaria, Chalara, Diplodia, Fusarium, Thielaviopsis*		(None known—possible natural immunity)	*Phytoplasma, Phytophthora*, (possibly *Mycoplasma*)
Grape	*Alternaria, Armillaria, Aspergillus, Botrytis*	*Agrobacterium, Xylella*	Grapevine viruses A, B, and D, tobacco mosaic virus, tomato ringspot virus	*Phytoplasma*
Legumes (beans, soya, lupin)	*Erysiphe, Fusarium, Sclerotium, Uromyces*	*Pseudomonas, Xanthomonas*	Bean common mosaic virus, bean golden mosaic virus, broad bean wilt virus	*Pythium*

(Continued)

Table 12.2 (*Continued*) Top 30 global crops with representative examples of associated diseases.

Crops in Alphabetical Order by Common Name	Disease Examples			
	Fungi	Bacteria	Viruses	Other
Maize	*Alternaria, Aspergillus, Botrytis, Curvularia, Diplodia, Glomerella*	*Clavibacter, Erwinia, Pseudomonas, Xanthomonas*	Barley yellow dwarf virus, cucumber mosaic virus, maize dwarf mosaic virus, maize streak virus, maize white leaf virus	*Mycoplasma, Phytophthora, Pythium, Sclerophthora, Spiroplasma*
Mango	*Alternaria, Aspergillus, Botrytis, Curvularia, Fusarium, Penicillium, Phoma, Septobasidium, Verticillium*	*Agrobacterium, Bacillus, Erwinia, Xanthomonas*	(Unknown virus in Pakistan mango crops, currently unidentified)	*Phytophthora*
Oats	*Claviceps, Erysiphe, Fusarium, Glomerella, Ustilago*	*Pseudomonas, Xanthomonas*	Oat necrotic mottle virus	*Pythium, Sclerophthora*
Onion	*Alternaria, Botrytis, Fusarium, Phoma, Rhizoctonia, Sclerotium, Urocystis*	*Pectobacterium, Pseudomonas*	Onion yellow spot virus, onion yellow dwarf virus	*Peronospora, Pythium*
Peanut	*Alternaria, Botrytis, Colletotrichum, Fusarium, Rhizoctonia, Sclerotinia*	*Pseudomonas*	Peanut clump virus, peanut mottle virus, peanut stunt virus	*Pythium*
Potato	*Alternaria, Botrytis, Choanephora, Fusarium, Puccinia, Septoria*	*Clavibacter, Pectobacterium, Pseudomonas, Streptomyces*	Potato leaf roll virus, potato virus X, potato virus Y, potato yellow dwarf virus, tobacco mosaic virus, tomato yellow mosaic virus	*Phytophthora, Pythium*
Rice	*Alternaria, Bipolaris, Curvularia, Fusarium*	*Burkholderia, Erwinia, Pseudomonas, Xanthomonas*	Rice ragged stunt virus, rice stripe virus, rice tungro spherical virus	*Pythium, Sclerophthora*
Rubber	*Colletotrichum, Corynespora, Ganoderma, Glomerella, Helicobasidium, Marasmius, Oidium, Rigidoporus, Ustulina*	*Agrobacterium, Bacillus*	None confirmed to date	*Cephaleuros virescens, Phytophthora, Pythium*
Sorghum	*Aspergillus, Colletotrichum, Fusarium, Penicillium, Sporisorium*	*Burkholderia, Pseudomonas, Xanthomonas*	Maize dwarf mosaic virus, sugarcane mosaic virus	*Phytoplasma, Phytophthora, Pythium*
Strawberry	*Alternaria, Botrytis, Fusarium, Glomerella, Mucor, Penicillium*	*Pseudomonas, Rhodococcus, Xanthomonas*	Strawberry mottle virus, strawberry vein-banding virus	*Phytoplasma, Phytophthora, Pythium*
Sugarcane	*Ceratocystis, Chalara, Fusarium, Gibberella, Puccinia*	*Herbaspirillum, Xanthomonas*	Sugarcane dwarf virus, sugarcane yellow leaf virus	*Phytophthora, Pythium*
Sunflower	*Alternaria, Botrytis, Fusarium, Gibberella, Macrophomina, Puccinia, Rhizopus, Verticillium*	*Agrobacterium, Erwinia, Pseudomonas*	Cucumber mosaic virus, sunflower chlorotic mottle virus, tobacco mosaic virus	*Phytophthora, Pythium, Plasmopara*
Tea	*Armillaria, Botrytis, Cercoseptoria, Exobasidium, Helicobasidium, Marasmius, Poria, Septobasidium, Xylaria*	*Agrobacterium, Pseudomonas, Xanthomonas*	Camellia virus 1	*Cephaleuros virescens*
Tomato	*Alternaria, Botrytis, Fusarium, Gibberella, Glomerella, Septoria, Stemphylium*	*Clavibacter, Erwinia, Pseudomonas, Ralstonia, Xanthomonas*	Tobacco mosaic virus, tomato bushy stunt virus, tomato spotted wilt virus	*Phytophthora, Pythium*
Wheat	*Alternaria, Aspergillus, Claviceps, Erysiphe, Fusarium, Gibberella, Glomerella, Puccinia, Rhizoctonia*	*Clavibacter, Erwinia, Pseudomonas, Xanthomonas*	Barley yellow dwarf virus, tobacco mosaic virus, wheat dwarf virus, wheat streak mosaic virus	*Phytophthora, Pythium*

This list is not exhaustive.
[a]*Canditatus* is a generalized genus-like name given to bacteria belonging to the family Rhizobiaceae that cannot be cultured and therefore cannot be ascribed to a specific genus.

best. The simple fact is that the far-reaching consequences of rapidly changing climatic conditions cannot be predicted. What we do know is that different weather conditions (for example, more rain or less rain, high or low temperatures, windy or calm conditions) all have an impact on crop growth, and also on the associated pathogens. In Chapter 4 we looked at the continually evolving arms race between pathogens and the developing set of host responses to these. With every change in climatic conditions, one partner in the pathogen–host relationship will be better suited to exploit the situation than the other. Plant breeders have a multi-faceted task on their hands. Not only do they have to develop crop plants with higher yields but less inputs ("smart crops") to meet ever-increasing demand, but also these plants need to be robust enough to survive new climates, and indeed overcome potential yield losses resulting from suboptimal conditions. One of the issues here is the simple fact that crops which have been bred over long periods have had an ever-decreasing gene pool to draw upon, so geneticists have had to go back to the beginning to look for desirable genetic traits in wild species relatives. Furthermore, as a result of increasing human populations, we have reached a stage where we are competing with *ourselves* for land to live on and land on which to grow our food.

Another potential outcome of climate change is a shift in the types of crops that can be grown in any given region. In the UK, the wine industry has flourished in recent years due to the warmer summers, and in the south of England some growers are already making inroads into new exotic crops normally associated with warmer areas, such as kiwifruits, olives, and chilis. These novel crop plants inevitably come with their own suite of associated pathogens that may or may not adapt to new conditions and transfer to new hosts. What is certain is that it is likely that we shall see more and more such changes to our farming landscape.

12.5 Precision Farming

Above we have discussed issues associated with changing climatic conditions. Although climate has clearly always been a dynamic and unpredictable factor affecting the growth of plants, especially crop plants, the rate and the more extreme ranges of climate change (for example, temperature and rainfall) are unprecedented in human history. Evolutionary responses may struggle to keep up, but farmers *must* keep up if they are to bring their crops to successful harvest. There are a number of technologies in their armory to enable them to cope, one of which is precision farming. This approach uses remote-sensing technology to observe the crop as a whole, analyze its requirements (for example, for nutrients, water, and pathogen and pest treatments), taking variability within the crop into account, and deliver exactly what is required, when and where it is needed. This reduces wastage and hence environmental damage caused by prophylactic use of unnecessary inputs, and it also ensures that yields are maximized.

12.6 Genetically Modified Organisms

The subject of genetically modified organisms (GMOs) is highly topical and controversial. Indeed it is a topic that generates emotional responses among many people who believe that companies which promote this technology are playing God. The alternative argument is that, in the face of an ever-increasing demand for food for our exponentially growing global population, we need every technology in our armory to tackle the threat of mass starvation. Here we briefly summarize the basic facts and leave our student readers to make up their own minds about the issue.

The rationale for genetically modifying crop plants is manifold. As discussed above, giving the host the edge in the host–pathogen arms race involves making a number of genetic improvements to the host plant. For example:

- The ability to resist disease and pests

- Resistance to herbicides, so that spraying kills off competitors but not the crop

- Improved tolerance of environmental pressures such as changing climatic conditions

- Increased fruit size and improved flavor

- Increased grain-fill (in cereals)

- Increased stem strength to prevent lodging (in cereals)

- Increased nutriment quality

- Increased secondary metabolite production (for medical or industrial purposes).

A discussion of the technology required to make these improvements is beyond the scope of this book, although information on techniques and applications is readily available in the literature and online. This includes such practices as gene silencing, inducing over-expression of transcription factors, chaperone-gene expression, binding-protein expression, and the introduction of artificial microRNA. Most of the research into the genetic improvements listed above is currently either restricted to the laboratory or is at the field trial stage and has not been rolled out to the wider environment—and indeed may never be, due to a myriad reasons, including concerns about human health, transmission of novel genetic traits to wild counterparts (with unknown consequences), and legislative restrictions making research into and development of new products non-viable with regard to long-term costs. Despite all of this, we cannot escape the fact that food production *must* be increased dramatically in order to meet current and future demands, and that strategies to meet these demands cannot be separated from developing a better understanding of food host plants and their associated pathogens at the molecular level.

12.7 Wheat

Due to modern breeding technologies, wheat yields have improved dramatically in the last 40 years, with yield productivity per hectare having more than doubled during that time. Globally, wheat production covers approximately 217 million hectares and accounts for approximately 20% of food calories. In June 2011, at the G20 conference of Agricultural Ministers from around the world, held in Paris, it was noted that despite yield improvements, for 6 out of the 10 previous years the demand for wheat had not been met. Furthermore, it is estimated that yields need to be increased by approximately 60–70% by 2050 if these demands are to be realized. This involves an approximate annual increase in production of 1.6–1.7%. One of the developments that emerged from this conference was the Wheat Initiative—a plan to focus on wheat production research, to improve communications between researchers, funders, and policy makers, to develop new varieties, and to develop better agronomic practices. At a follow-up conference in 2015 in Turkey, a comprehensive agenda aimed at achieving these aims was developed. Wheat is typically beset by a multitude of diseases, and yield losses resulting from disease can be up to 20% annually. At the Turkey conference, the Wheat Initiative committed to the following strategies: the recording of ongoing genomic-based pathogen surveillance and epidemiology; the identification and characterization of novel sources of resistance to the main wheat diseases over the next 5 years; the identification of associated mechanisms within 10 years; and the deployment of resistance strategies resulting from this research in the longer term.

12.8 Summary

Earlier in this book we described how students often regard agriculture and horticulture as "boring" and old-fashioned, but it is clear from all that we have discussed in this text, and indeed from the available literature, that quite the opposite is true. From the latest genetic studies through to mechanical and computer technology this topic is right up to date, as it needs to be if we are to tackle food security issues and bring food to the table—for everyone. Each year, global deaths from famine are in the hundreds of thousands, and ongoing hunger affects millions. There are over 7 billion people currently living on our planet, of whom approximately one in eight are going hungry. Meanwhile populations continue to increase. Studies in plant pathology are at the cutting edge of scientific research in a broad range of fields, and it is imperative that this subject area attracts new recruits to take these technologies forward—the very existence of our species may depend upon it.

Saving our planet, lifting people out of poverty, advancing economic growth ... these are one and the same fight. We must connect the dots between climate change, water scarcity, energy shortages, global health, food security and women's empowerment. Solutions to one problem must be solutions for all.

(Ban Ki-moon, Secretary General of the United Nations, January 2007 to December 2016)

The food security of a populous country is more than just economics and trade. A drop of 1% of China's grain output means extra imports of nearly 5 million tons or 2.5% of the world's total grain trade volume.

(Dr Cheng Guoqiang, Secretary General and Senior Research Fellow, Academic Committee, Development Research Center of the State Council of China)

Further Reading

Acquaah G (2012) Principles of Plant Genetics and Breeding, 2nd ed. Wiley-Blackwell.

Agrios GN (2005) Plant Pathology, 5th ed. Elsevier Academic Press.

Dunwell JM (2014) Transgenic cereals: current status and future prospects. *J Cereal Sci* 59:419–434.

Mba C, Guimaraes EP & Ghosh K (2012) Re-orienting crop improvement for the changing climatic conditions of the 21st century. *Agric Food Security* 1:7.

Wills R, McGlasson B, Graham D & Joyce D (2007) Postharvest: An Introduction to the Physiology and Handling of Fruit, Vegetables and Ornamentals, 5th ed. CABI Publishing.

Useful Websites

www.fera.defra.gov.uk/plants/ planthealth/imports

Global Food Security programme. www.foodsecurity.ac.uk

Wheat Initiative. www.wheatinitiative.org

Glossary

abiotic Factors that influence biological systems, but organic in origin i.e. physical or inorganic chemical factors

adenosine triphosphate (ATP) The coenzyme molecule responsible for temporary storage of energy

allele One of two paired or alternative forms of the same or equivalent gene

amphiploid A hybrid organism with at least one full set of diploid chromosomes from each parent

apothecium (pl. **apothecia**) In ascomycete fungi, a cup-shaped fruiting body

appressorium (pl. **appressoria**) The specialized flattened parasitic fungal cell structure that develops from the germ tube of a fungal spore and puts pressure on the plant host prior to penetration

arbuscular (also called **endomycorrhizal**) Relating to fungi whose mycorrhiza penetrate the cortical cells within the host plant's roots and there form highly branched structures called arbuscules

argonaute protein A protein that has a major role in the RNA silencing complex

ascospore In ascomycete fungi, any of the sexually produced spores that develop within the ascus

ascus (pl. **asci**) In ascomycete fungi, the structure within which the ascospores develop

bine In certain climbing plants, particularly hops, a stem that spirals around a supporting structure

binomial system The system of nomenclature used to name all living organisms, consisting of the genus and the species

biofilm One or more species of microorganism within a polysaccharide matrix that is produced by those microorganisms

biotroph A plant pathogen that remains associated with its host for an extended period and does not kill it

boom and bust The theory that a pathogen and its host are alternately successful in defenses and counter-defenses

canker A lesion, usually in woody tissue, that oozes fluid exudate; the term is sometimes used to refer to certain fungal and bacterial diseases

capsid The protein shell that surrounds the nucleic acid of a virus

capsule In bacteria, a sticky layer of protein or polysaccharide surrounding the bacterial cell wall

chemoautotrophic Relating to an organism that gains its nutrients from inorganic sources

chlorosis Loss of green pigmentation due to a reduction in the amount of chlorophyll in plant tissues

clade A group of organisms that share the same features because they are all derived from a common ancestor

coenocytic hyphae Hyphae that lack septa

conidia (sing. **conidium**) Asexually produced, non-motile fungal spores

conjugation Non-sexual sharing of genetic material by transfer of plasmid DNA from one bacterium to another

cryptic species Morphologically identical organisms that belong to different species and cannot interbreed

cutin A polymer, composed of fatty acid molecules, that is a component of the plant cuticle

cytoplasmic streaming The directional flow of cytoplasm within plant or fungal cells, allowing the transport of materials such as nutrients

damping off The collapse and death of seedlings resulting from infection by any of a number of different fungal pathogens

dehiscence The spontaneous and "explosive" release of spores from fungi, or of seed from plants

Deuteromycota Generalized term for the imperfect fungi, which do not fit into any taxonomic group; they are also known as mitosporic fungi, because they only reproduce by mitosis

dikaryotic In fungi, relating to a cell unit or compartment containing a pair of compatible nuclei

dipterocarp Any member of a family of tropical trees found mainly in lowland forests of South-East Asia

DNA deoxyribonucleic acid, which is the double-stranded polymer that contains encoded genetic information; it is located in the nucleus, mitochondria, and plastids

domain The highest level of taxonomic grouping

drupe Any stone fruit (for example, plum, peach, cherry)

ectomycorrhizal Relating to fungi whose hyphae do not penetrate the cortical cells of the host plant's roots, but merely surround the root; such fungi have a mutualistic relationship with the host

effector molecule A molecule produced by a pathogen that binds to proteins of the host plant, thereby suppressing the host defense mechanisms

elicitor A molecule on the surface of a pathogen that induces host plant defense mechanisms

endophyte Any microbe that for at least part of its life cycle lives within a plant in a mutualistic relationship

epidemiology The study of disease spread throughout a population

Eumycota The kingdom consisting of the true fungi

exopolysaccharide A polysaccharide that is produced and secreted into the environment by microbes

expression The physical manifestation of genetic coding by production of a gene's product (RNA or protein)

fastidious Selective

fimbria (pl. fimbriae) Any of the fine proteinaceous hair-like structures on the surface of bacteria, used for adhesion

Five Kingdom Classification System The classification of living organisms into five kingdoms, a taxonomic system that was extensively used until recent developments in taxonomy

germ tube A fungal outgrowth from a spore that ultimately develops into a new hypha

glumes In grasses, the pair of membranous bracts that surround the spikelet

Gram staining A staining method that is used to distinguish between two groups of bacteria, namely Gram-positive bacteria (which have a thick peptidoglycan layer) and Gram-negative bacteria (which have a thin peptidoglycan layer)

green bridge Overwintering crop debris and weeds that may contain a range of disease inoculum which can infect subsequent crops

guttation Exudation of droplets of xylem fluid via glands called hydrothodes from plant surfaces (mainly the leaves) as a result of root pressure

haustorium (pl. haustoria) In a parasitic fungus, the specialized hyphal structure that is used to extract nutrients from the plant host after it has penetrated the plant cell

hemibiotroph A pathogen that can utilize both parasitic and saproxylic nutrition, as in the initial stages of infection the plant host cells remain alive, but as the disease progresses they collapse and die

heteromorphic Relating to any organism that has two or more physical forms during its life cycle

hexaploid Containing six homologous sets of chromosomes

hypha (pl. hyphae) Any of the long thread-like filaments of fungal mycelia

hyphal rhizoid A small branching hyphal structure

ice nucleation The triggering of super-cooled liquid water to freeze

inclusion body Any of the small structures found inside viral host tissue, generally composed of viral aggregates, viral coat proteins, and additional helper proteins

infection peg A small outgrowth of the appressorium that penetrates the host plant

integrated pest management (IPM) A method of integrating conventional pest management practices with cultural and organically based practices, resulting in a maximally effective management strategy

lenticel In plants, a pore that traverses the woody tissue, such as bark, enabling gaseous exchange to occur

lichen A close mutualistic association between a fungus and an alga, the structure and nomenclature of which imply that it is a single organism

lignin A non-carbohydrate polymer that is a structural component of plant cells, providing rigidity. It is especially abundant in woody plants

lodging At the internodes, the tendency of a plant stem to bend over so that it is no longer vertical. It may interfere with normal growth and development of the plant and decrease harvesting efficiency

masting The synchronous production of large numbers of seed, especially by certain tree species (for example, those in the family Dipterocarpaceae)

mesosome An invagination in the plasma membrane of a bacterium

mitosporic fungi An alternative term for the imperfect fungi (that is, those that do not fit into any taxonomic group and only reproduce by mitosis); they are also known as the Deuteromycota

mollicute A bacterium-like microbe that lacks cell walls

monocyclic Relating to a pathogen that has a single infection cycle per season

mutualistic Relating to a symbiotic relationship between at least two organisms, from which all parties benefit

mycelium (pl. **mycelia**) A mass of thread-like hyphae

necrotroph A pathogen that kills the host tissue as it invades it, resulting in the death of all or part of the host organism. The pathogen extracts nutrients from the decaying material

non-caducous spores Spores that detach from the sporangium for dispersal

nucleoid The region in a bacterial cell that contains most of the genetic material but is not bound by a membrane

null mutant A mutant copy of a gene that does not carry out that gene's normal function

ontogenic resistance Changes in the level of host resistance to disease susceptibility with age

opine Any of the compounds induced in gall tumors and then used by pathogenic bacteria as a nitrogen source

paraphysis (pl. **paraphyses**) Any of the filamentous support structures present in some fungal species

pathogen Any microbial organism that causes disease in another organism (the host) by living at its expense

pathotype A disease-causing variant of a microorganism

pathovar A pathogen strain that is a genetic variant, varying only in the host that it infects

peptidoglycan The polymer layer that lies outside the plasma membrane in bacteria

periplasmic space In Gram-negative bacteria, the space between the inner and outer plasma membranes

phagocytosis The process by which a small particle or material becomes engulfed by a cell

photoautotrophic Relating to an organism that obtains nutrients by converting inorganic compounds into organic compounds (most commonly by photosynthesis)

photosynthate Any product of photosynthesis

phyllomorphy The abnormal development of the floral parts of a plant into prolific leafy growth, typically as a result of infection by phytoplasmids or viruses

phylloplane The surface of a plant leaf that provides a habitat for microorganisms

phylogeny The evolutionary history of an organism, often depicted like a family tree showing how the organism is related to other species

phylum (pl. **phyla**) The taxonomic grouping that is intermediate between kingdom and class

phytoalexins Anti-microbial compound produced by plants in direct response to pathogen challenge

phytoanticipin Any of a group of constitutive defensive secondary metabolites produced by plants

phytophagous insects Insects that feed on plants

pilus (pl. **pili**) A tube-like appendage on the surface of bacteria, used mainly for attachment to other cells, and for the exchange of genetic material during conjugation

plasmid In bacteria, a small circular DNA molecule that can replicate independently of the chromosomes, and that can be transferred from one organism to another

plasmodesmata Any of the microscopic channels that cross the cell walls between plant cells, and allow transport of materials from cell to cell

pleiotropic Relating to a single gene that affects more than one phenotypic trait

pleomorphic Relating to variation in size and shape of cells of the same type, particularly the ability of some bacterial cells to alter their size or shape in response to changes in environmental conditions

polycyclic Relating to a pathogen that has several infection cycles per season

polyketide A biologically active secondary metabolite produced by a plant that enhances its defenses

prokaryote An organism in which the genetic material in the form of DNA is located in the cytoplasm, not in a membrane-bound nucleus. Prokaryotes belong to the domains Bacteria and Archaea

Protista A taxonomic kingdom consisting of single-celled organisms

protomer Any of the repeating copies of proteins that make up the structure of a viral capsid

quorum sensing The autoinduction system of host defense-resistant bacterial signaling molecules

race An informal taxonomic grouping that is intermediate between subspecies and strain

rachis In cereals such as wheat, the main supporting axis of an inflorescence

reverse transcription (RT) The enzyme-catalyzed process that enables the production of viral DNA from a viral RNA template

rhizosphere Soil region in close proximity to plant roots and thus influenced by root exudates and associated microbes

ribosome Any of the small cellular organelles that are the site of protein synthesis in the cell

RNA Ribonucleic acid, one of the structural components of genetic material

root depletion zone The zone surrounding the roots where there is a lower concentration of mineral nutrients, due to their uptake by the roots

roguing The removal of diseased plants and the plants immediately surrounding them

saponin Any of the plant-derived glycosides that have "foaming" characteristics

saprophytic Relating to an organism, especially a fungus, that lives on dead and decaying material

sclerotia (sing. **sclerotium**) In fungi, hardened mycelial resting bodies containing food reserves that enable them to remain dormant for long periods

secondary host A temporary host used by a pathogen for a short transition period before it returns to the primary host; unlike primary hosts, secondary hosts usually display no pathogenic symptoms

secondary metabolites Any metabolic product that is not essential for cell growth and development, or for reproduction, but that has a specialized secondary role, especially in protection or signaling (for example, antibiotics, pigments, signaling compounds)

secondary thickening The increase in girth of a plant that occurs as a result of lateral meristematic activity

septum (pl. **septa**) In fungi, any of the cross-walls within the cells that effectively create separate cell units

sporulation In some bacterial species, a dormant state involving the formation of hardened "spores" that can withstand adverse conditions for long periods

stomastyle Alternative name for a stylet

Stramenopiles The taxonomic kingdom that includes the water molds and downy mildews

stroma (pl. **stromata**) In fungi, a dense structural tissue that produces fruiting bodies

stylet In aphids and some nematodes, a component of the mouthparts that is formed from four tubes and which is adapted for piercing the cell wall of leaf phloem cells of the plant host, thereby providing access to nutrients for feeding

suberin A waxy waterproof substance that is found in the cell walls of some types of plant tissues

suckering In many trees and shrubs, vegetative reproduction involving the development of new shoot growth from the roots

symbiotic Relating to a close and mutually beneficial relationship between two or more organisms of different species

syncytium A continuous mass of cytoplasm containing many nuclei, formed as a result of the merging of many cells

syringomycin A cytotoxin produced by many strains of the bacterium *Pseudomonas syringae*

taxonomy The study of the classification of living organisms

tiller In grasses, any of the stems produced after the development of the initial seed leaf

tribe The taxonomic grouping that is intermediate between family or subfamily and genus

urediniospore Any of the spores produced in the uredium of rusts

vector Any living organism that transfers infectious agents from one primary host to another

virion A complete virus particle

viroid A plant pathogen that is smaller than a virus and consists only of a small circular single-stranded RNA molecule

wettable powder A pesticide or fungicide formulation in which the active ingredients are ground into a fine powder to which a wetting agent has been added, for ease of application in the form of a liquid spray

white collar complex A light perception complex that has a fundamental role in signaling pathogenesis by fungi

Woronin body In ascomycete fungi, a peroxisome-derived "plug" that blocks septal pores to prevent loss of cell contents

zoospore A flagellated asexual spore

zygospore The diploid reproductive spore of fungi and some fungus-like organisms

Index

Figure 1.4 Morphology of various infectious agents on plants. (a) Gray mold (*Botrytis cinerea*) on fruit of raspberry (*Rubus* species); (b) brown rot (*Monilinia* species) on fruit of apple (*Malus* species); (c) mildew on leaf of greater plantain (*Plantago major*); (d) leaf spot (*Mycosphaerella fragariae*) on leaves of strawberry (*Fragaria* species); (e) black rot (*Xanthamonas* species) on brassica leaf.

Figure 1.9 Infection (a) rust on onion leaves; and (b) mildew on wheat leaves.

Figure 1.13 Rust pustules on leaves of wheat (*Triticum* species).

Figure 2.13 Crazy top disease (*Sclerophthora macrospora*) is the cause of unusual morphological changes in plants such as maize (*Zea mays*). (Courtesy of FX Schubiger, pflanzenkranheiten.ch)

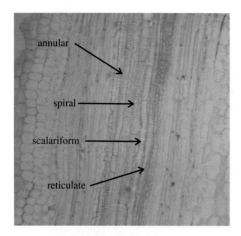

Figure 3.3 Examples of secondary thickening tissue types in a eudicotyledenous stem.

Labels on figure: annular, spiral, scalariform, reticulate

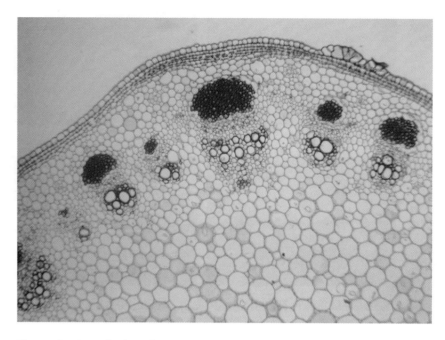

Figure 3.4 Vascular bundle tissue from a eudicotyledenous stem.

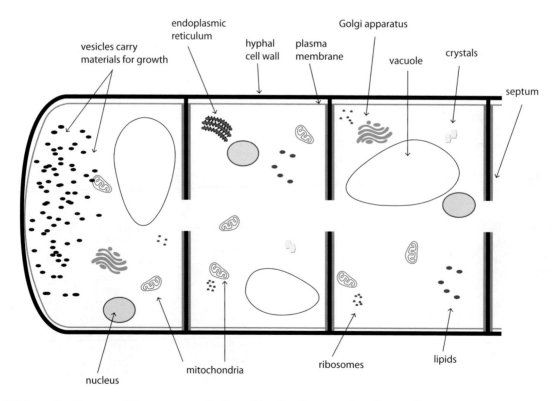

Labels on figure: vesicles carry materials for growth, endoplasmic reticulum, hyphal cell wall, plasma membrane, Golgi apparatus, vacuole, crystals, septum, nucleus, mitochondria, ribosomes, lipids

Figure 3.5 Schematic diagram of the structure of a fungal hypha. Growth takes place at the tip, and the materials required are transported to the tip in vesicles. Note that the cell compartments are only partially separated by the septa, thereby allowing easy movement of these materials.

Figure 3.13 Mycelia of powdery mildew on strawberry (*Fragaria* species).

Figure 3.16 Tulip bloom showing the characteristic morphology of color-break virus.

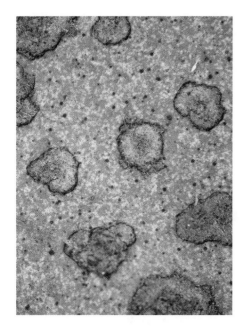

Figure 3.17 The algal plant parasite *Cephaleuros virescens* on leaf of *Citrus* species. (Courtesy of Cesar Calderon, USDA APHIS PPQ. Bugwood.org.)

Figure 3.19 The holoparasitic plant dodder (*Cuscata* species).

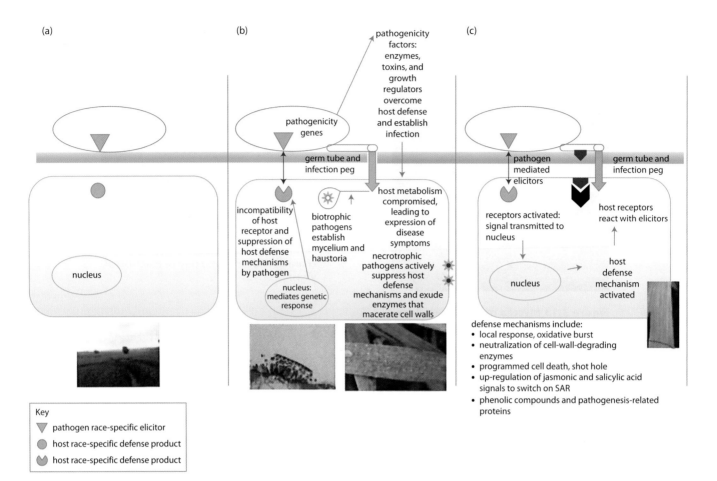

Figure 4.1 Schematic diagram of infection and detection and host response. (a) There is no recognition reaction, so the pathogen does not recognize the host species, and there is no infection. (b) There is a compatible reaction between the host species and the pathogen, leading to germination of the pathogen spore, suppression of the host's defense mechanisms, and successful infection. (c) There is a recognition reaction in which the host has a suite of defense mechanisms to suppress infection by the pathogen. The pathogen may germinate, but successful establishment and subsequent infection are suppressed by the host. SAR, systemic acquired resistance.

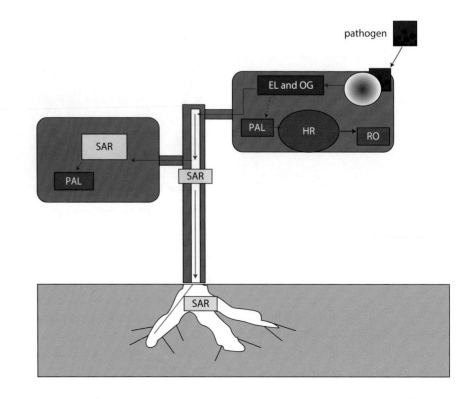

Figure 4.3 Schematic diagram of systemic acquired resistance. R, recognition (for example, gene for gene); EL, elicitors; OG, oligogalacturonides; HR, hypersensitive response; RO, reactive oxygen; PAL, phytoalexins; SAR, systemic acquired resistance.

Figure 6.4 Powdery mildew on (a) sweet pea leaf, (b) courgette leaf, (c) oak leaf, and (d) clematis bloom.

Figure 6.26 Orange rust (*Puccinia kuehnii*) on sugarcane. (Courtesy of Josiane Takassaki, Ferrari – Laboratório de Doenças Fúngicasem Horticultura, Instituto Biológico.)

Figure 6.28 Rhododendron bud blast, showing the brown to black bud and the white pinhead coremia, which contain conidia.

Figure 6.30 Distinct black tar-like lesions characteristic of acer tar spot (*Rhytisma acerinum*). (Courtesy of Andrew Richards, formerly of Plymouth University.)

Figure 6.34 Apple scab (*Venturia inaequalis*) on fruits, showing (a) a mild infection, (b) a moderate infection, and (c) a severe infection in which the apples have completely disintegrated as a result of scab infection and susceptibility to secondary environmental damage caused by sunburn.

Figure 6.35 The effect of apple scab (*Venturia inaequalis*) on a susceptible variety of Braeburn compared with a resistant variety. A heavy infestation of the disease in the susceptible variety is visible both (a) in the whole tree and (b) in a selection of the fruits from the same tree. The resistant variety of Braeburn shows a complete lack of infection in both (c) the whole tree and (d) the leaves and fruit from the same tree.

Figure 7.2 *Erwinia amylovora* (fire blight) is a common bacterial disease of trees and shrubs, especially species belonging to the family Rosaceae, such as apples (Malus species shown here), pears, hawthorn, and many others. (Courtesy of Robert L Anderson, formerly of USDA Forest Service, Bugwood.org)

Figure 8.7 Tobacco mosaic virus (TMV) on sunflower. (a) Collection of sunflowers: note that the specimen on the right is much smaller, in terms of both height and leaf and bloom size, due to infection by TMV; (b) the infected sunflower shown in more detail; (c) one of the leaves of the infected sunflower—note its small size and distorted, leathery, and mottled appearance.

Figure 9.2 *Phytophthora* crown rot on pod of cacao (*Theobroma cacao*). (Courtesy of Christopher J Saunders, Bugwood.org)

Figure 9.1 Field-scale image of potato crop infected with *Phytophthora infestans*.

Figure 9.7 *Phytophthora ramorum* infection of sweet chestnut (*Castanea sativa*).

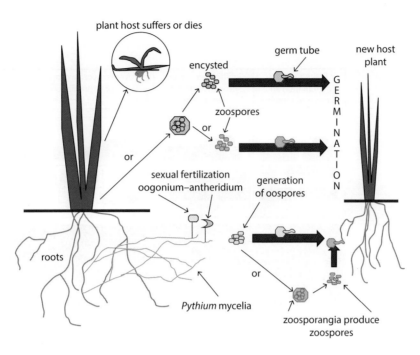

Figure 9.9 Schematic diagram of the life cycle of the family Pythiaceae using a generalized plant host indicating both sexual and asexual reproduction.

Figure 10.1 Tractor-mounted spraying equipment.

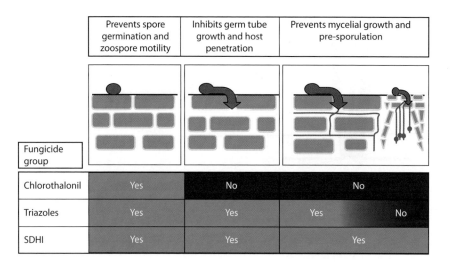

Fungicide group	Prevents spore germination and zoospore motility	Inhibits germ tube growth and host penetration	Prevents mycelial growth and pre-sporulation	
Chlorothalonil	Yes	No	No	
Triazoles	Yes	Yes	Yes	No
SDHI	Yes	Yes	Yes	

Research Box 10.1 Figure 2 Schematic representation of disease control by key fungicides on fungal phases of the pathogen life cycle. SDHI, succinate dehydrogenase inhibitors.

Figure 11.1 A range of soil fertility-building crops. (b) Vetch.

Figure 11.5 Weeds within a red cabbage crop. The weeds can act as secondary hosts of many pathogens.

Figure 11.6 Leek rusts. (b) Close-up view of rust pustules on leek leaf.

(a)   (b)

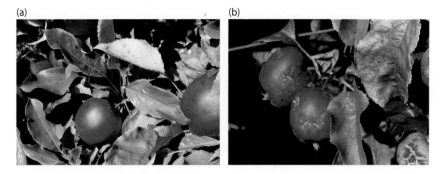

Figure 12.1 The Braeburn apple (*Malus domestica* var. "Braeburn") is a popular variety sold in the supermarkets. (a) shows resistant cultivar and (b) shows susceptible cultivar.